Methods in Field Geology

F. Moseley

University of Birmingham

Methods in Field Geology

W. H. FREEMAN AND COMPANY

Oxford and San Francisco

W. H. Freeman and Company Limited
20 Beaumont Street, Oxford, OX1 2NQ
660 Market Street, San Francisco, California 94104

Library of Congress Cataloguing in Publication Data
Moseley, Frank, 1922–
Methods in field geology.
Includes index.
1. Geology—Field work. I. Title.
QE45.M67 550'.723 80–25472
ISBN 0–7167–1293–8 (US)
ISBN 0–7167–1294–6 (US: pbk)

Set by Western Printing Services Ltd, Bristol, England
Printed in the United States of America

Preface

Fieldwork and geological survey form important parts of undergraduate courses in the geological sciences, but these topics are covered only cursorily in standard textbooks for geology students. This book is an attempt to fill what is, I believe, a serious gap in the literature.

In Part I survey methods for students are reviewed. My examples are taken largely from North-Western Europe, but there are also some from Spain, Cyprus, Africa, Arabia and Greenland. The principles illustrated apply to surveys in most parts of the world. This is important because geologists, perhaps more than people in other professions, are required to work in a wide variety of climates and terrains. It is also important to appreciate the contrasting techniques required for different types of rock and structure: metamorphic, volcanic and sedimentary rocks have to be surveyed differently, and Quaternary deposits—much more important than many university courses suggest—also require special treatment.

In Part II I have selected case histories from various locations to illustrate in greater detail the survey principles outlined in Part I. Among the cases are: surveys on sedimentary, volcanic and metamorphic rocks, and on glacial deposits in Britain; structural mapping in Spain; general reconnaissance in Oman, Libya, Kenya and Greenland; Hydrogeological mapping in South Yemen; and mapping of coastal sections in Cyprus. Although I deal with specific locations in Part II, I have indicated examples of analogous regions where the same techniques are applicable.

This book is a result of many years' experience in field geology, but it is important to acknowledge the firm foundations provided by the teaching of Professor F. W. Shotton and the ideas contributed by colleagues and friends at the University of Birmingham and elsewhere. I am also grateful to Miss Sonia Hodges who prepared a large proportion of the text figures in the Drawing Office at the University of Birmingham; to the Royal Air Force and Royal Engineers, without whose support none of the Arabian and African surveys would have been possible; to the Governments of Oman, Spain, Libya, Kenya and Denmark for permission to work in their territories; and finally to my wife who has put up with long absences on fieldwork and has criticized some of my more extreme thoughts.

F. M.

Contents

Illustrations are by the author unless otherwise stated

Methods in Field Geology

PART I General Principles

1 Introduction

Field geology has always formed a basis for other geological activities; geological maps and field reports are used not only for specific studies such as geochemistry and geophysics, but by civil engineers, mining and oil companies, town and country planners and many others, not forgetting the amateur geologist who simply wishes to know more about the surrounding countryside. It is therefore of the greatest importance that geological maps, plans, measured sections and other data be accurate within stated limits, so that users can rely on the information presented. If there is uncertainty its nature should be made clear either on the map itself or in any written account which may accompany it. To these ends the training of a field geologist should aim at objectivity, and flights of imagination, so tempting and at times valuable, should be reserved for special pages of a field notebook. This 'objective' information should be recorded on field maps and in field notes in such a way that others can take the documents into the field, locate the exposures listed and get the same answers to lithologies, geological boundaries, structural data and so on. Perhaps this will seem like a statement of the obvious, but anyone who has tried to work with the field notebooks of other geologists will take the point. It is a common experience to find data recorded in such a random and chaotic manner that after a lapse of time not even the original observer can interpret the results satisfactorily. When such is the case the ultimate publication or report cannot fail to suffer. There is of course a place for imagination and intuition in field geology, but it cannot be stated too strongly that this facet should not be interpolated with what, for want of a better expression, may be called the factual observation of a field map and notebook.

A geological investigation may have a variety of purposes, and civil engineers and oil companies, to name only two, will require different sets of facts. The most comprehensive type of survey is usually that undertaken by Government geological surveys because they aim to cover all aspects and produce a service to all possible users; whereas oil geologists, engineering geologists, university research workers and others will generally restrict themselves to more specialized maps and reports related to their particular requirements and interests.

The variety of field geology is enormous and far beyond the scope of any one person, as it includes a large number of activities in rocks of all ages and in very different geographical regions that range from tropical forests to arctic mountains, with the requirement varying from detailed plans to reconnaissance maps. Field geologists must therefore vary their approach to the subject according to circumstances. Most have their own methods of operating and recording. No one method is necessarily better than any other and readers should realize that the suggestions here reflect to a considerable extent my personal preferences. Large organizations exploring for oil, coal or minerals, where success yields great financial rewards, will use expensive methods beyond the scope of any research worker: boreholes may be placed on a 50-metre grid, geophysical and geochemical surveys may blanket an area of economic interest, and specialists in a dozen or more fields may be asked for their opinions. There are also major parts of the Earth's surface, especially the oceans, where new methods of exploration are being continuously evolved, but which are not considered in this account.

This is not the first book to describe methods of field geology. There have been many others, perhaps the most comprehensive being that of Compton (1962), a magnificent work which deals with the subject from first principles onwards. This new book is not intended to replace Compton but to complement it. The more elementary concepts are dealt with as briefly as possible, but some of the practical details arising from them are expanded. The intention is to describe field methods that can be applied by final year undergraduates; some of the suggestions should also be useful to postgraduate research workers. Most of the subjects covered during the first year of geology at university will be 'taken as read'. It will be assumed, for example, that readers can use a compass/clinometer, that they can identify common rock types in the field using a hand lens, that they will know something about recording field notes and plotting geological lines on maps. They should also be familiar with the elements of topographical survey and photogeology; these are described fully by Compton.

2 Equipment

There are two categories of equipment used by field geologists: (1) general equipment most of which would normally be carried and (2) more specialized equipment applicable only to certain types of survey.

General equipment

Hammers. Sledge, 1 pound ($\frac{1}{2}$ kg) or $2\frac{1}{2}$ pound (1 to $1\frac{1}{2}$ kg) depending on the type of rock being investigated.

Chisel.

Tape measures. 2-, 10- or 30-metre tapes depending on the task.

Hand lens. × 10 or 15 areal magnification is necessary.

Compass-clinometer. A liquid immersion variety is desirable.

Dilute acid. For carbonate tests.

Collecting bags and marking pens. For the numerous specimens that must be collected if the fieldwork is to be supported by subsequent laboratory studies; for example, for some structural and palaeomagnetic studies orientated specimens are necessary (see chapter 16).

Map case (Fig. 1). This should be large enough to take aerial photographs, overlapped so that they can be inspected stereoscopically in the field. A suitable case can easily be made from hard plastic, by taking three sheets of approximately 40 × 20 cm and hinging them along the long edge to form a four-page book. This will take aerial photographs, field maps, field notes and other information that it may be necessary to carry. Leather straps can be attached using rivets, and a combination of elastic bands, bulldog clips and adhesive tape used to hold the papers in place.

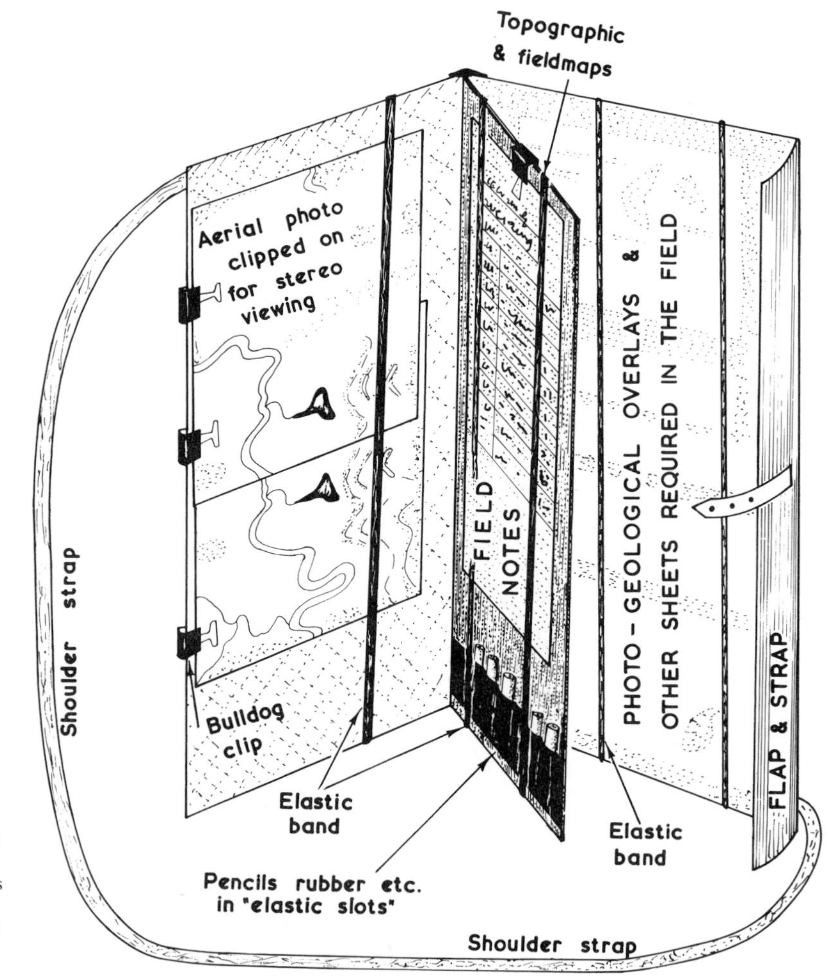

Fig. 1 A pattern for a geological mapping case with four pages to contain maps, aerial photographs and field notes. If aerial photographs are to be clipped in for stereo-viewing a convenient size is 40 × 27 cm.

Aerial photographs. These should be sufficient for the day's fieldwork and to cover the views of distant hillsides. Topographical maps are often inadequate and geological data may have to be plotted on to aerial photographs. This can be done either directly on to the photographs using all-purpose pencils, on to photocopies of the photographs or on to transparent overlays placed on the photographs. The last method allows more detail to be recorded. Unless one is fortunate to have more than one copy of each photograph, no permanent marks should be made on them, since this obscures photographic features and decreases objectivity. A suitable solvent can be carried to remove all-purpose (or chinagraph) pencil marks when it becomes necessary.

Pocket stereoscope (Fig. 2). This is essential for most field surveys, not only for locating position on aerial photographs, but also for the geological information which can be seen and plotted during mapping. Geologists who have used stereoscopes for these purposes will confirm that they add a new dimension to the accuracy, speed of mapping and interpretation. Although some geologists can see aerial photographs stereoscopically without the use of a stereoscope, most, especially beginners, cannot. Pocket stereoscopes can be bought at reasonable cost, and it is also possible to buy cheap lenses (focal length 8 cm) and construct a home-made stereoscope. I use two such lenses mounted as goggles and fastened round the back of the head with elastic. They are easier to 'slot into place' and use in the field than a stereoscope with 'legs', which has to be used on a horizontal surface.

Fig. 2 Pocket stereoscopes. The normal use, shown on the left, is when the aerial photographs can be placed on a table or similar flat, horizontal surface. Field use will often be under more adverse conditions, perhaps in wind and while standing on a steep hillside. It is then better to adapt the stereoscope as spectacles (right) when both hands will be available for holding the map case and for writing. Some geologists are able to see stereoscopically without using a stereoscope—an ability well worth acquiring.

Cameras. Geologists should try to buy a good camera early in their career. It is an advantage to have one with interchangeable lenses (for example, with focal lengths of 28 mm, 50 mm and 135 mm and extension tubes). A polaroid camera is also useful. Students should consider buying a good basic camera and adding interchangeable lenses as funds permit. Remember that the weight of a camera and its accessories—and all other field equipment—is considerable; a survey of a quarry with a vehicle available is a different matter from survey of an inaccessible mountain area.

Clothing. The importance of adequate clothing and some elementary precautions necessary on geological survey must be stressed. The items below apply particularly if the survey is to be conducted in a highland region, where extreme climatic conditions (cold) may be experienced, but remember that such conditions are often found in relatively mild countries such as Britain even in summer. Fig. 3, taken in October, illustrates this point clearly.

1. Waterproof anorak, overtrousers and spare warm clothing should be carried. Thin trousers can be dangerous, even lethal, to anyone who is wet and subjected to exposure.
2. Boots with commando-type soles should normally be worn, but gum boots may be appropriate when stream sections or extremely boggy areas are being examined.
3. Gaiters are useful in wet vegetation and essential in snow.
4. In addition to a compass and a map, a whistle, torch, small first aid kit, spare high calorific food and a survival bag should be taken.
5. The route for the day should be pinned to the tent, stuck to the car or left at lodgings.
6. If the survey is in the tropics or the Arctic different criteria apply; these are discussed in chapters 14 and 16 to 19.

Fig. 3 A photograph of a field party in the Cumbrian mountains, northern England, in late October. Weather conditions in such areas can be extremely changeable. On this particular day the morning in the valley was mild, but a snowstorm and high wind developing towards lunchtime resulted in arctic conditions on the mountain. It is wise to be prepared for such events.

Specialized equipment

Augers. For sampling unconsolidated deposits (for example, Quaternary formations) an auger is usually used. For straightforward mapping a lightweight, small-diameter screw auger (say 3 cm) is sufficient, but a large-diameter type (15 or 40 cm) will be necessary for detailed sampling. The type will depend upon the purity of sample required and on the lithology which is to be penetrated (Figs 85 and 86).

Aneroid barometer. This can be invaluable, particularly when mapping relatively unknown mountainous terrain in a reconnaissance style. Fairly reliable altitudes can be obtained provided there are frequent checks with a base camp reading (see chapter 18).

Surveying apparatus (level, plane table, theodolite, etc.). This may be required for detailed surveys.

Binoculars. These are particularly useful in mountainous areas for picking out detailed structures on distant hillsides, which might otherwise be missed.

Microscope. A binocular microscope at base camp can be particularly useful for examining specimens collected during the day.

Equipment for extreme conditions. Expeditions to high mountains, particularly in the Arctic and Antarctic, to desert regions and tropical rain forests will require much specialized equipment. Information on some of the requirements is given in Part II.

3 Initial preparation

General

It is obvious that maximum information should be obtained before venturing into the field: publications referring to the area to be investigated and descriptions of analogous areas and analogous problems elsewhere should be consulted, aerial photographs should be studied and a photogeological map prepared. The time a geologist can spend in the field is invariably limited and since the object must be to achieve the maximum result in the time available, initial preparation is of paramount importance; the working model of a photogeological map is perhaps the most important of all.

Photogeological maps

There are several books on photogeological methods (e.g. Allum 1966; Lattman and Ray 1965; Miller 1961) and here it is necessary only to summarize procedures, which vary considerably depending on the scale of the aerial photographs and on the intended detail of the mapping. In the United Kingdom the photographic scale is usually 1:7000 to 1:10 000. Elsewhere 1:25 000 is common, but some regions are covered at only 1:40 000 to 1:80 000 and there are still large areas without any air photograph cover. Oil companies and other multinational organizations have areas specially flown at the scale and with the type of photographs (black and white, colour, false colour, infra-red, etc.) best suited to the survey. Most field geologists, however, including most research workers and probably all undergraduates, have limited funds, and they must make do with whatever they can obtain; some suggestions are considered here.

One method is to use all-purpose pencils directly on the aerial photographs; this is more suitable for reconnaissance survey since detail cannot be plotted. These can then be combined with photocopies of the photographs, on to which detail is transferred using a hard pencil. Possibly the best method is to record geological details on a transparent overlay placed over the aerial photograph. If the area is particularly important enlargements of the photographs (plus photocopies) can be used. Reconnaissance mapping can be considered as scales of 1:25 000 or less, and detailed mapping as scales of from 1:10 000 to 1:2500.

When plotting geological data avoid the margins of the photographs since here there will be distortion of both scale and direction, particularly in mountainous terrain (Fig. 4). The plotted information should include lineaments and differences in the texture and tone of the photographs (Figs 7 and 13). These will represent geological boundaries, faults, joints, cleavage, drift features and a variety of lithologies; the significance of these may not be fully understood at this stage but this will become apparent as the ground survey gets under way. In addition, topographical features such as escarpments, gullies, streams, ridges and hill tops should be recorded, partly so that the photogeological map can be matched with the topographical map, and partly because some of these features may have been plotted inaccurately on the topographical maps, and will require correction. In spite of the stress on objectivity it is advisable to introduce some interpretation into a photogeological map—for example, differentiation between bedding, joints and cleavage and between solid and drift—otherwise the map can become a meaningless hotchpotch of lines. Ways of doing this are as follows:

1. Coloured inks can be used for solid geology, drift deposits and topography (say black, red and green).

2. Solid geological information can be used to distinguish different structures; for example, bedding features can be dotted lines, faults and joints can be firm lines (note that this has not been done on Figs 7 and 9). Fig. 9 should be compared with Fig. 8 and with the subsequently published map (Fig. 10).

Fig. 4 The effect of steep ground on the apparent orientation of structures seen on aerial photographs. Photogeological maps of part of the Lilloise Massif, East Greenland, showing gently dipping basalt lavas intruded by basic dykes inclined 75° west. The zigzag line represents an arête rising to between 600 and 900 m above the glacier. The centre points of photographs 1, 2 and 3 are situated on the west of the ridge, in the centre and on the east of the ridge respectively, and it will be noticed how the apparent orientation of the dykes differs from one photograph to another.

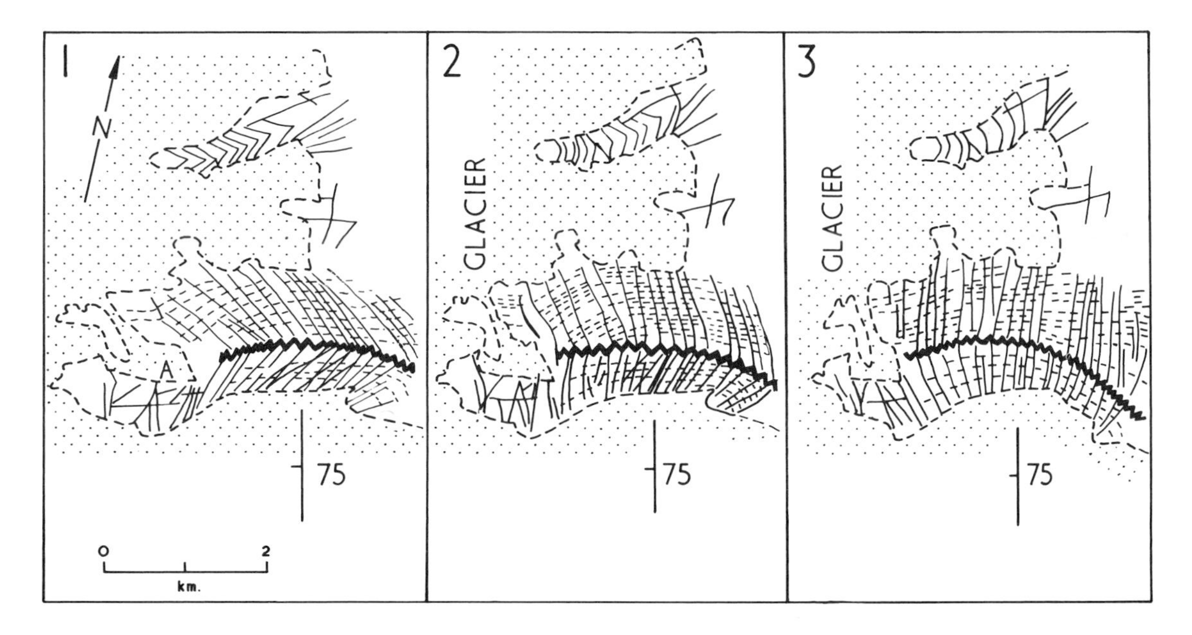

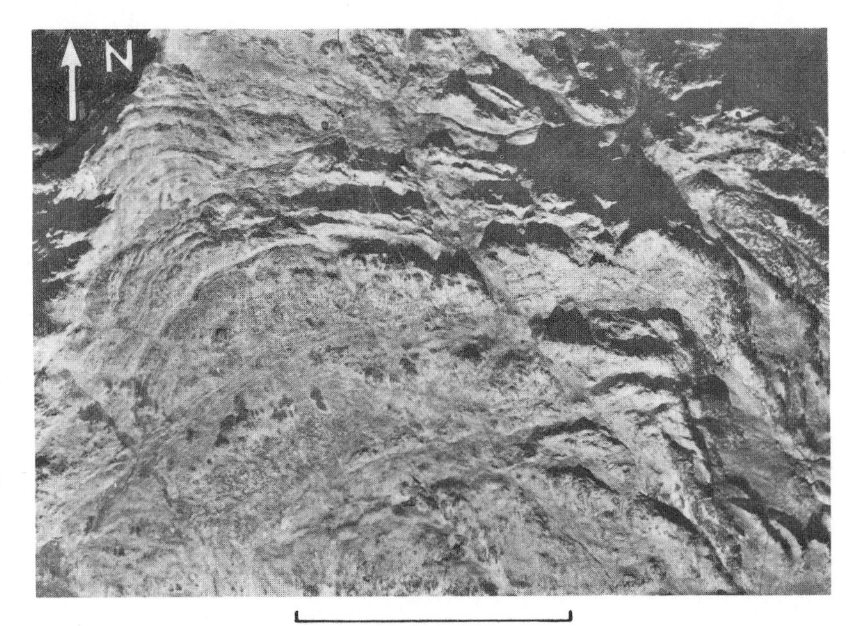

Fig. 5 An aerial photograph of the High Spy area, Borrowdale, English Lake District. The outcrops of rock formations and positions of faults are clearly seen. Use of a stereoscope made it possible to construct the photogeological map of Fig. 7, which was scale-corrected as indicated by Fig. 6. Field traverses at a later date showed that the succession consisted of alternations of andesite lava and tuff. (Crown copyright/RAF photograph.)

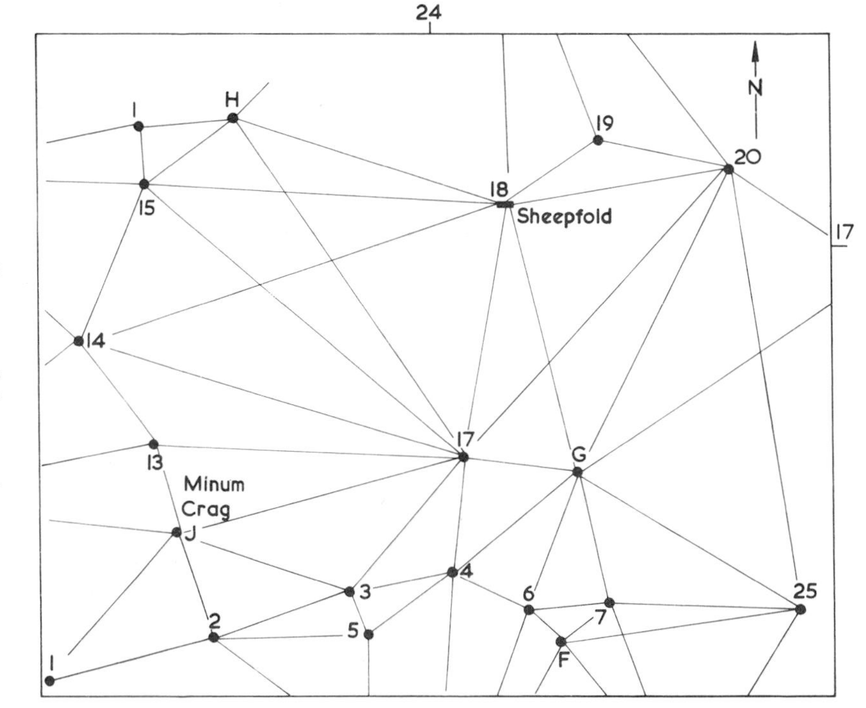

Fig. 6 Scale-correction of a photogeological map when an accurate topographical map is available. There will not be a constant scale from one part of an aerial photograph to another. When the ground is flat there will be scale differences between the centre and the edges of the photographs, and scale differences are more pronounced where there are altitude differences from one part of a photograph to another (Fig. 16). This diagram shows a simple method of scale correction, and refers to Figs 5 and 7. A number of localities which are easily identified on both photograph and topographical map are connected by lines into a triangular grid which makes it easy to transfer the geology, triangle by triangle, from a photogeological map to a scale-correct map.

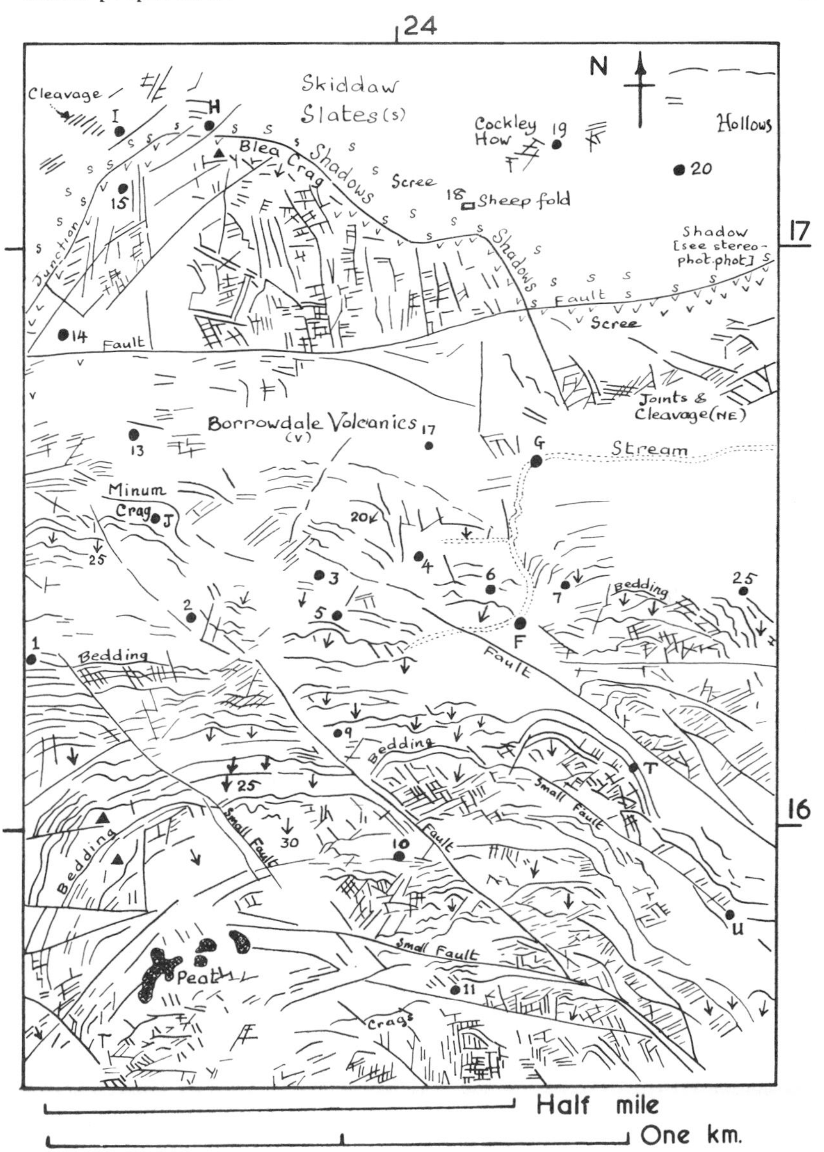

Fig. 7 Scale-corrected photogeological map of High Spy, Borrowdale, English Lake District (see Figs 5 and 6). Such a map, prepared prior to field survey, should include a certain degree of provisional interpretation. In this case the bedding is clearly visible over much of the area, and estimates of dip are fairly accurate, but it was not possible to give reliable interpretations of lithology (alternations of lava and tuff). Major faults are easily seen but their displacements are uncertain, and a number of the shorter lines are more likely to be master joints. From general knowledge of adjacent regions it was suspected that, of the short lines, the NW and N trends are joints and the NE trends represent cleavage.

The numbers and letters are localities (sheep fold, bend in stream, etc.) which are accurately marked on 6-inch maps and can also be seen on the aerial photographs. They facilitate the scale-correction of the photogeological map (Fig. 6). In practice different coloured inks would be used for topography, solid geology, drift, etc.

Fig. 8 Aerial photograph of part of the Bannisdale Slate outcrop, NW England. Compare with Fig. 9 (compiled from stereophotographs) and Fig. 10. (Crown copyright/RAF photograph.)

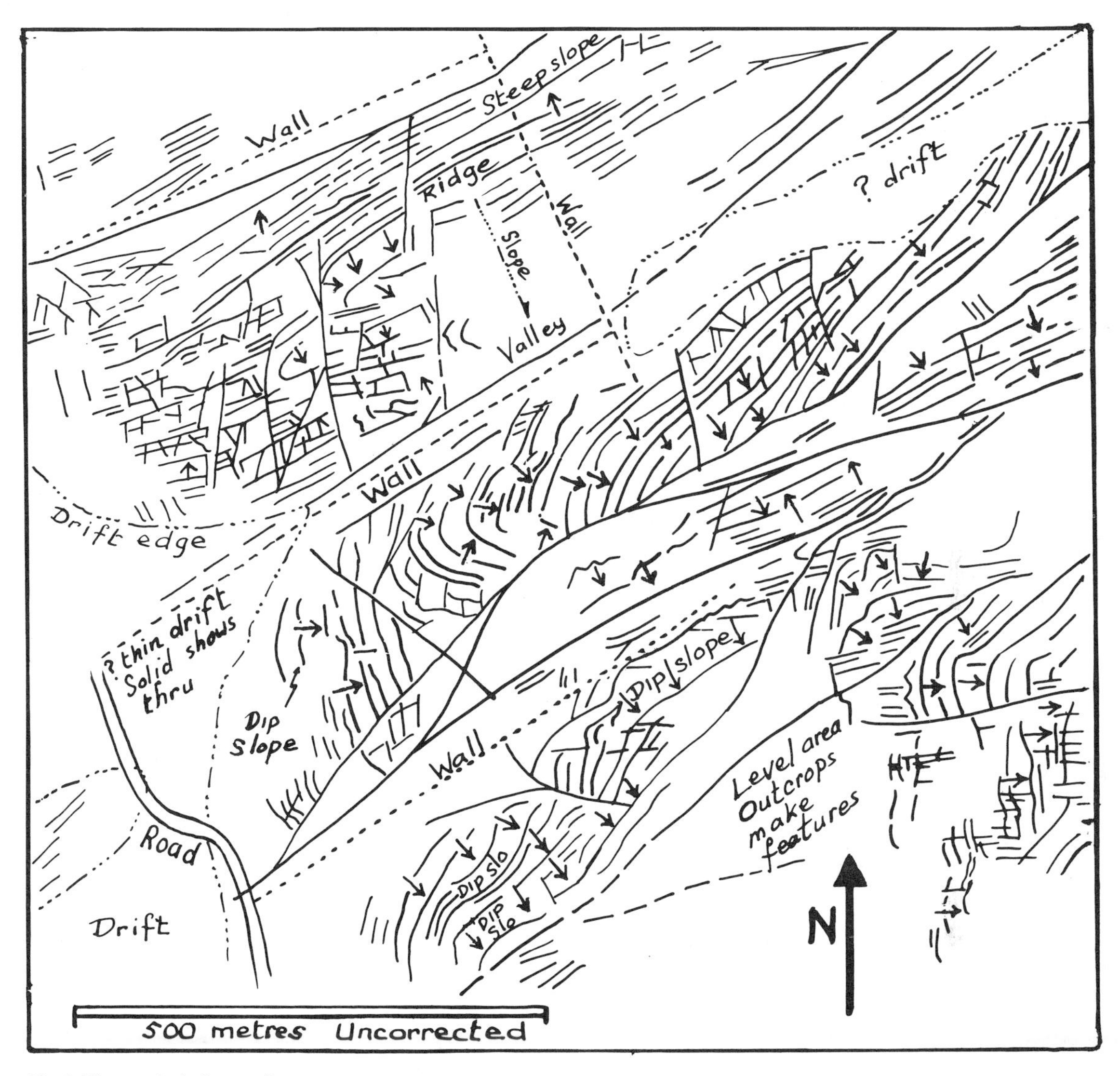

Fig. 9 Photogeological map of part of the Bannisdale Slate outcrop, Silurian, NW England. On this map both bedding and faults are shown by continuous lines. The short lineaments are joints and cleavage. Compare this map with Figs 8 and 10.

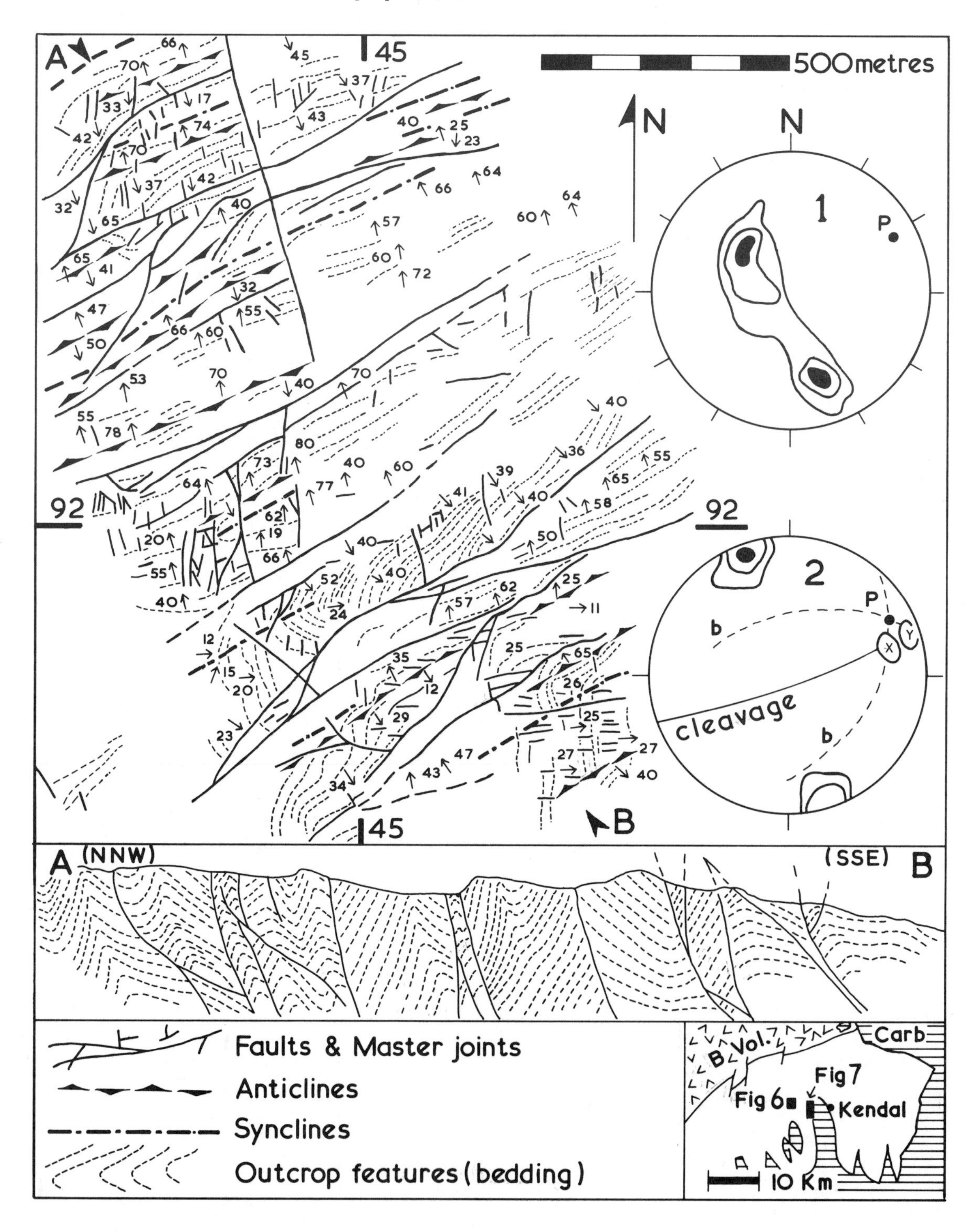
500metres
N
N
1
P
2
P
b
cleavage
b
X
Y
92
92
45
45
A
B
A (NNW)
(SSE) B
Faults & Master joints
Anticlines
Synclines
Outcrop features (bedding)
B. Vol.
Carb
Fig 7
Fig 6
Kendal
10 Km

◁ *Fig. 10* Published map following photogeology and field investigation of the Bannisdale Slates near Crook, NW England (Moseley 1972a). Field measurements of bedding dip are shown on the map, and totals of bedding and cleavage orientation are shown on the stereographic projections. It is interesting that the 6-inch to the mile field map of the original survey 100 years ago (Aveline *et al.* 1888), was almost identical to this map and was made very quickly, during what would now be classed as reconnaissance survey. (© Geological Society of London.)

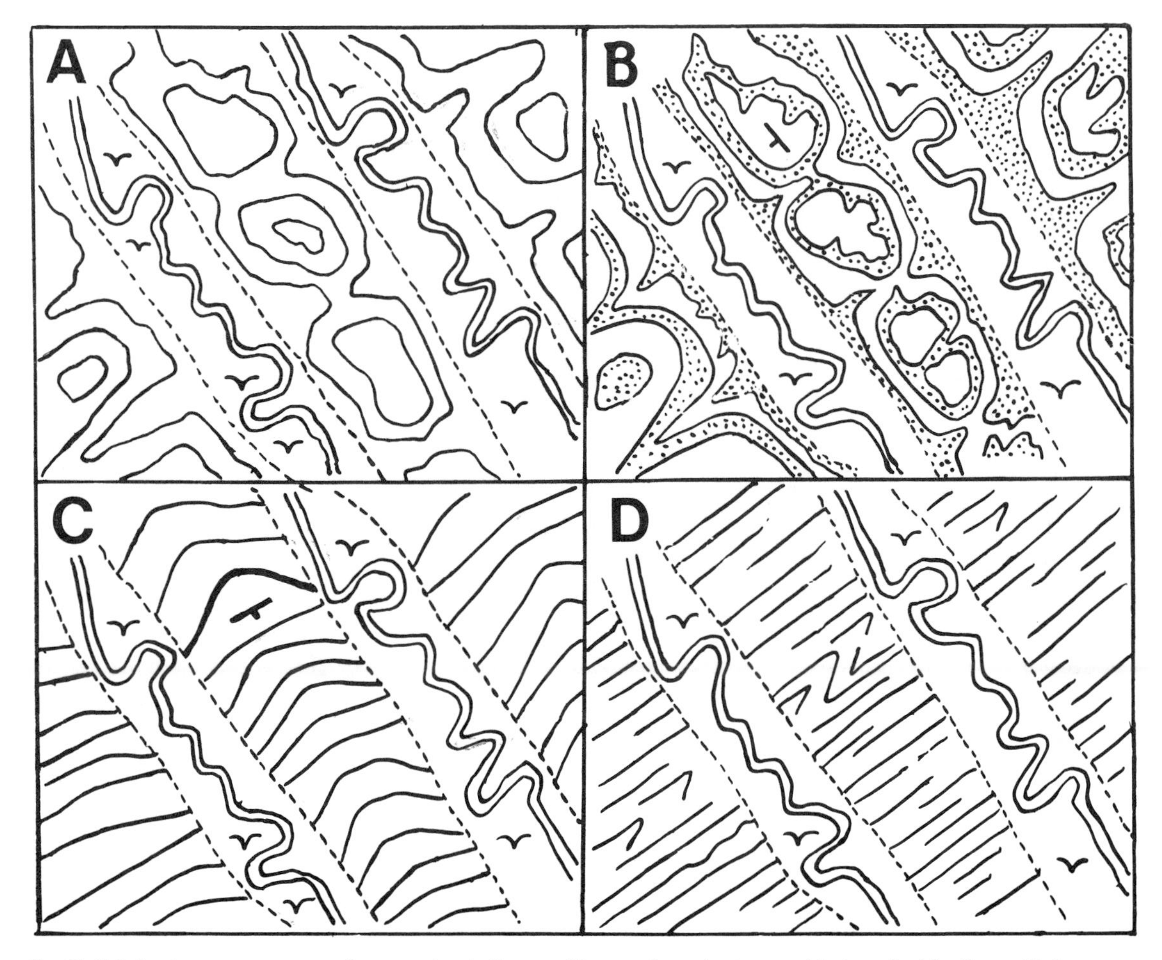

Fig. 11 Relation between structure and topography. **A**, Topographic map of a region traversed by two alluvial valleys, with the topography represented by contours. **B**, **C** and **D** show the outcrop patterns of layered rocks (e.g. sedimentary or volcanic sequences), which can be seen and interpreted at a glance on both aerial photographs and geological maps. **B**, Horizontal strata; **C**, steeply inclined strata; **D**, vertical folded strata.

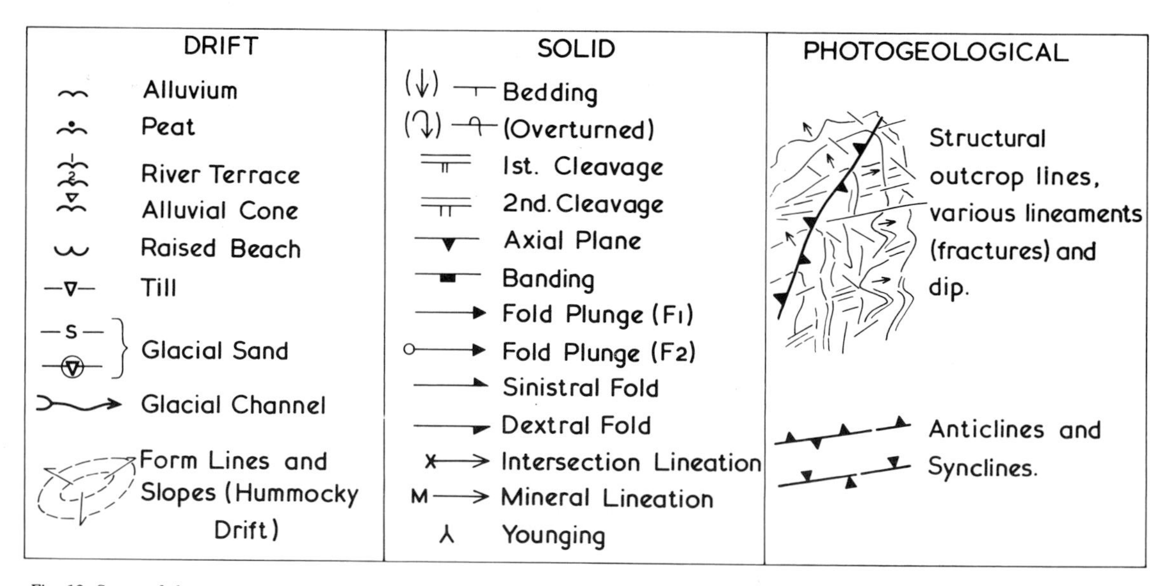

Fig. 12 Some of the geological symbols commonly used on field maps.

3. Dip of bedding can be represented under the categories of gentle, moderate or steep (say less than 10°, 10 to 40° and more than 40°). The inclination of bedding can be determined more exactly should this be required. It is preferable to use conventional arrows for dip, rather than the strike and dip symbol, since the latter can become confused with short lineaments representing joints (Figs 7, 9 and 12).

4. The lithology of the rock is perhaps the most difficult to estimate, but provisional attempts can be made. Most obvious is the relation between topography and alternations of hard and soft rock, and almost equally obvious are the contrasting tones and structural patterns of such rocks as quartzite, granite, basalt, limestone, etc. (Figs 12 and 13). Suitable ornament can be devised for these clear-cut rock types and should be in the form of crayon shading or stipple. A line ornament will become confused with lineaments. There will always be a limit to the correct and detailed identification of rocks, and the final stages must be left to the field survey.

5. Drift features, such as alluvium, terrace, peat and moraine, should also be distinguished. Form lines can indicate the shape of morainic mounds; dotted lines can represent scree; the slope of gravel fans can be shown (Figs 27, 67, 83 and 84).

6. Topography (ridges, summits, gullies, escarpments, etc.) can be represented by approximate form lines, particularly where the topographical map is inaccurate or generalized, or when a geological boundary is controlled by topographical detail.

Fig. 13 (opposite) Mosaic of part of Hallaniya, Kuria Muria Isles, Oman. The tones of the rocks together with their structures permit provisional photogeological interpretation of: L, limestone; S, predominantly serpentinite; C, probable layered complex, ultra mafic to mafic; G, granite; D, dolerite dykes; F, fault. The lithologies are clear on this photograph, but are very much more obvious on the original photographs when viewed stereoscopically. Five photographs were used in this mosaic. (Crown copyright/RAF photographs.)

The final result—a photogeological map which looks like a geological map—will provide an extremely valuable basis for the field survey to follow. It is also important that a geologist should be able to interpret the general structure at a glance as illustrated by Fig. 11.

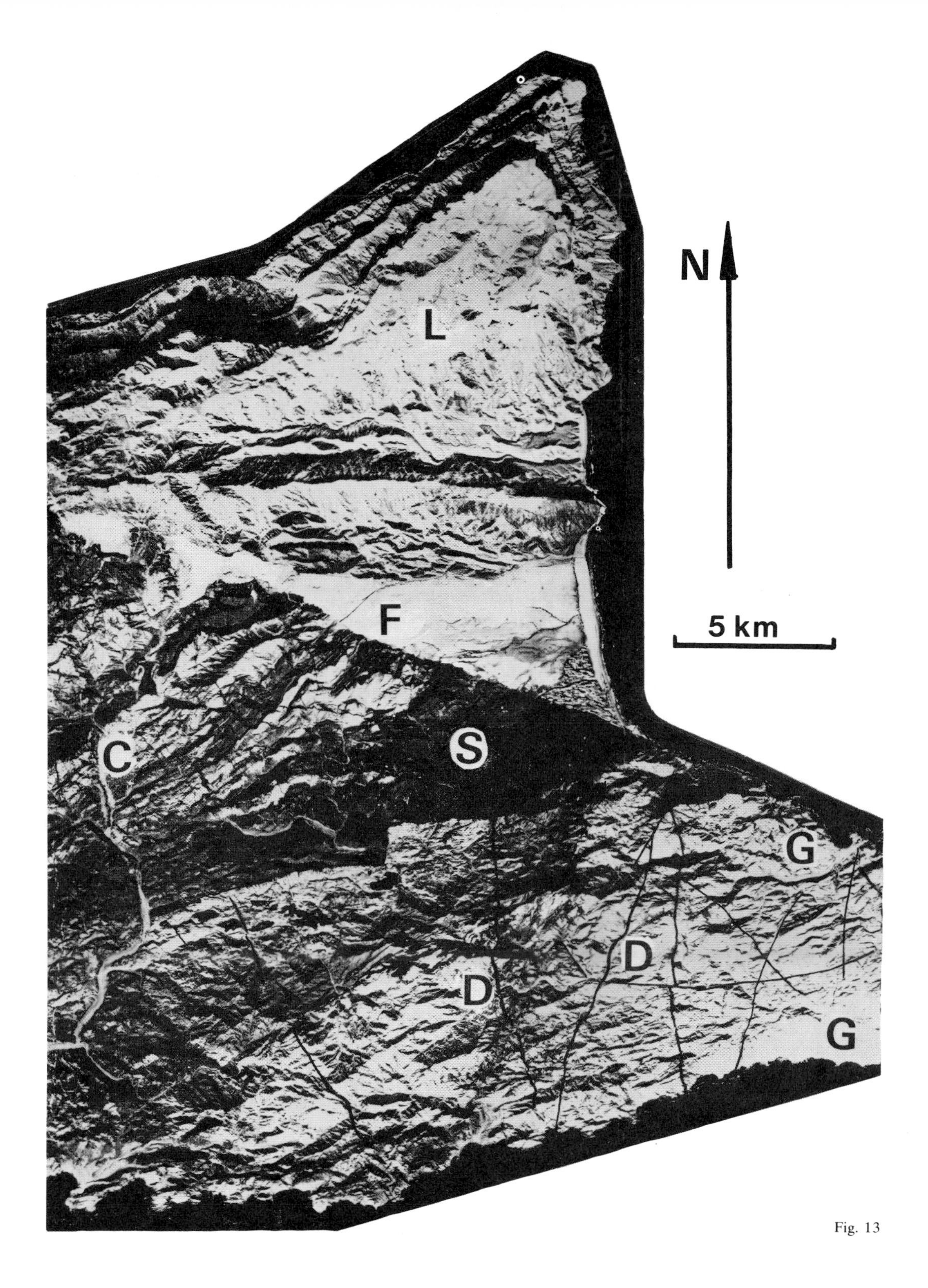

Fig. 13

4 Use of aerial photographs in the field

It was suggested in chapter 2 that a map case designed for holding stereopairs of aerial photographs and a pocket stereoscope should be carried as essential pieces of field equipment. These will give a three-dimensional model of the ground that is always available for inspection. Not only does it permit identification of localities with far greater precision than is possible using most maps, but it also enhances geological features, the significance of which may not be apparent to a ground observer. For example, in mountainous country it may at first be confusing to find rough crags and gullies rising on all sides without any obvious pattern; a glance at the aerial photograph model, however, may reveal unsuspected simplicity, and this can save a great deal of time.

In most cases aerial photographs (and/or photocopies) with transparent overlays can be used as field maps, with geological data recorded on them in the same way as indicated in chapter 3 (page 13). If much detail is required it is better to obtain enlargements of the photographs and these (or photocopies of them) can be used as field maps and data recorded directly on to them. However, in the more highly populated areas of developed countries good quality large-scale maps with extensive topographical detail (roads, streams, field boundaries, buildings, etc.), may be available and it may then be quicker and more efficient to plot information directly on to the map rather than on to aerial photographs as additional time is required to transfer data from photograph to map. When aerial photographs are used for plotting field data, it is tempting but a mistake to use the prepared photogeological map. If this is done the map becomes too cluttered; it is better to start a new map for the field data and to carry the photogeological map in the map case for reference (observe, for example, the detail on Figs 7 and 9). Other methods which may be preferred or necessary in special cases are described in Part II, especially in chapters 14 and 16.

The procedure which then follows is more or less common to all geological mapping projects. Geological boundaries and lithologies plotted on photogeological maps are confirmed, modified or corrected on field maps, and the significance of other photogeological lineaments (joints, faults, cleavage, dykes, etc.) is determined. Structural data and locality numbers are also recorded in their correct positions.

5 Compilation of geological field maps from aerial photographs and field observations

As a field survey progresses it is important that field data recorded on aerial photographs, and the confirmed photogeological observations are transferred to accurate field maps. These may be large-scale published maps (1:10 000 etc.); if there are no suitable base maps available they may have to be prepared from the aerial photographs.

Transfer of data to reliable maps

Rescaling will always be necessary, even when aerial photographs and map are at the same scale, since there will be distortion of the former away from the centre. Orientations of linear features in mountainous areas are particularly liable to be misleading (Fig. 4) since the camera is almost certain to view them obliquely. Similarly outcrops on steep hillsides can be either foreshortened or extended, and scales on mountain tops will be larger than those in the valleys. Allowances must be made for all these factors and scale corrections made (Figs 4 and 16).

The simplest general methods for transferring data from photographs to maps have been adequately described elsewhere (e.g. Allum 1966; Lattman and Ray 1965). If there is sufficient topographical detail already plotted on a base map it is easy to compare the photographs with it and transfer the geological information to the correct positions. On Fig. 7 the numbers and letters refer to points on the base map which are clearly recognized on the photographs. If a change of scale is necessary a sketch master will save time, but it is easy to improvise as indicated on Fig. 6.

Transfer of data to unreliable maps and preparation of maps from aerial photographs

The various ways of making geological maps from aerial photographs with little or no ground control have been well described elsewhere and so only the general methods are summarized here. The procedure is to compile either a print laydown or a mosaic (Allum 1966, pp. 7 and 81), which will be either semi-controlled where there has previously been some topographical survey, or uncontrolled where the region is unsurveyed. In the latter case a limited topographical survey would be

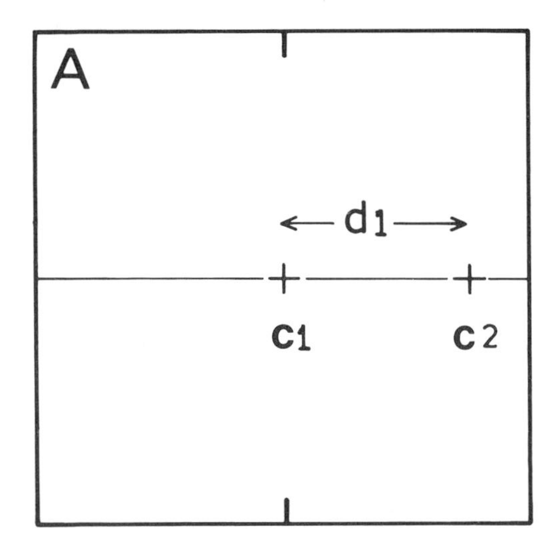

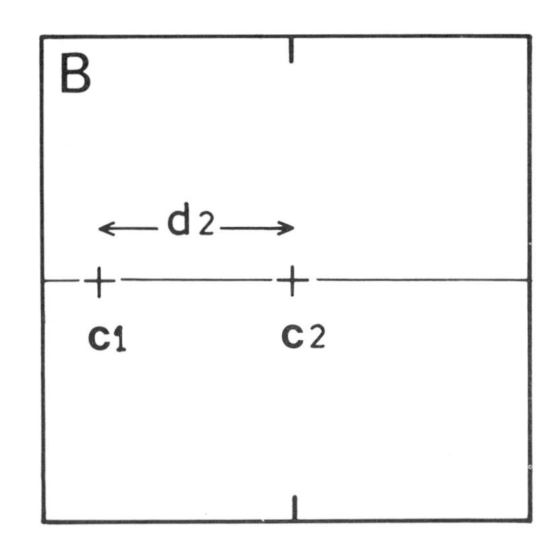

Fig. 14 A pair of stereoscopic photographs showing the centre or principle points (C1 and C2), and the air base or photo base (d1 and d2). An average of d1 and d2 is taken if they are of slightly different length. C1–C2 is the flight line.

highly desirable as part of a geological survey. The general method is first to establish a form of triangulation using the centre points of the aerial photographs to draw a base line, followed by the construction of templets or azimuth lines so that, first, a line overlap of one air photo run can be assembled (Fig. 15) and then, if the sidelap has been correctly flown, adjacent runs can be fitted together (Allum 1966, p. 86; Lattman and Ray 1965, p. 101). An accurate map can quickly be made using this method, but the type of equipment for optimum results is unlikely to be available to undergraduates, who will probably have to make do with the more laborious but still quite effective hand templet method. This involves marking control points as shown on Fig. 15. Continuous lines connect the centre points of adjacent photographs, and radial lines are drawn to other control points, selected so that they are easily recognizable elements of the topography; a prominent bend in a river for example. These points will make it possible to compile a line overlap and to relate it to adjacent ones, although the latter step is rather more difficult because of scale distortion towards the photograph margins. This can be particularly awkward in mountainous terrains where the radial lines are unlikely to intersect at a point, but if there is some topographical control, reasonably accurate maps acceptable for geological reconnaissance can be drawn. The geological maps of Masirah, Oman (Moseley 1969a; Moseley and Abbotts 1979) and Beihan and Fahdli (Moseley 1971a and b), both in South Yemen, were constructed in this way and are the subject of case histories described in Part II.

Fig. 15 (opposite) Construction of hand templets for the assembly of line overlaps and mosaics. **A** shows the centre (principle) points, the transposed centre points (1, 2, 3) and the wing points, which are obvious topographic features. Continuous lines join the centre points and short radial lines are drawn through the wing points. **B** shows the assembly of the three photographs into a line overlap and **C** shows how line overlaps can be assembled into a mosaic or print-lay-down.

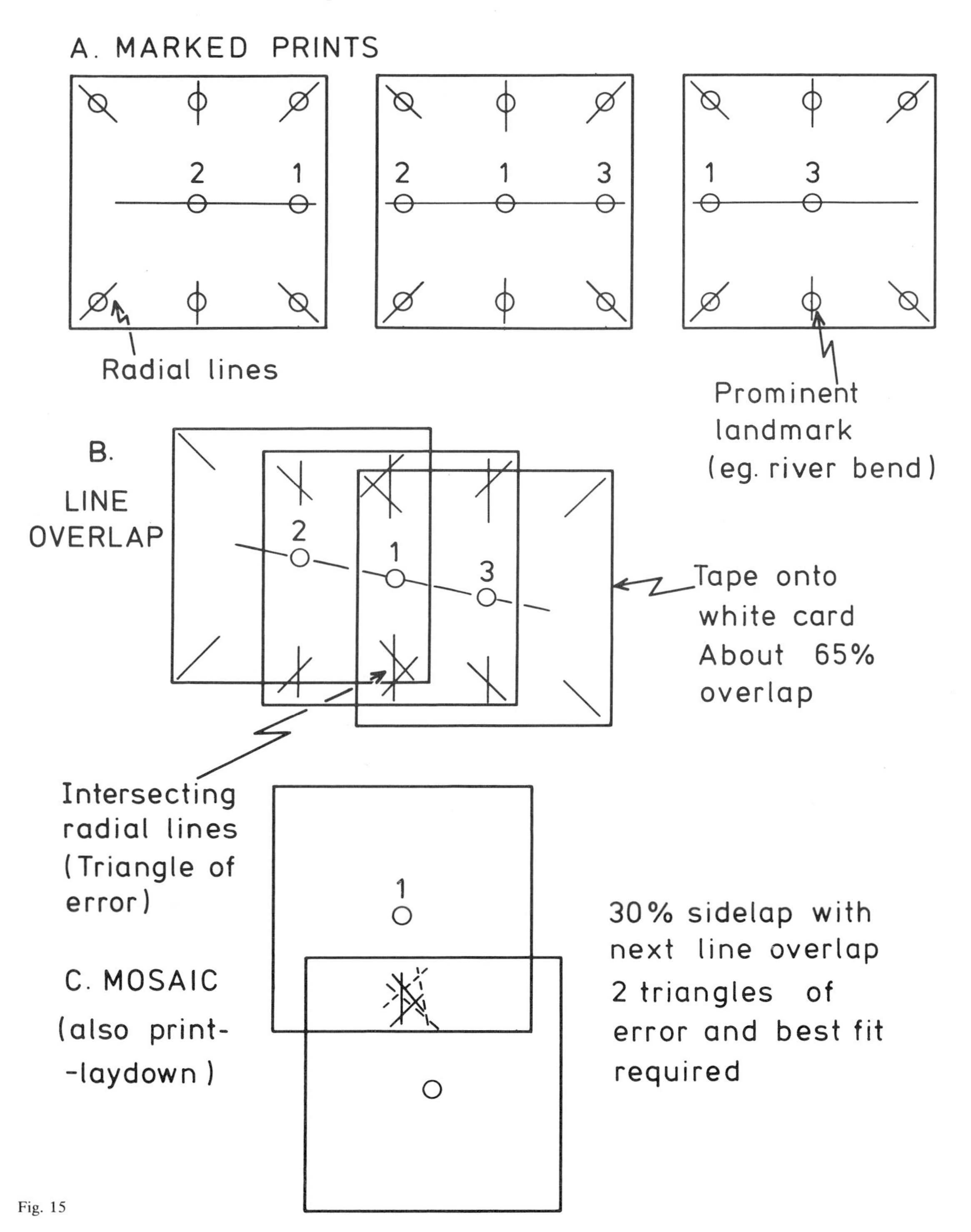
A. MARKED PRINTS
2
1
2
1
3
1
3
Radial lines
Prominent landmark (eg. river bend)
B.
LINE OVERLAP
2
1
3
Tape onto white card About 65% overlap
Intersecting radial lines (Triangle of error)
1
C. MOSAIC
(also print--laydown)
30% sidelap with next line overlap
2 triangles of error and best fit required

Fig. 15

Determination of scale and height on reconnaissance maps from aerial photographs

Scale

Determination of scale from aerial photograph information alone is unreliable unless the photographs have been flown with this purpose in mind. Nevertheless there are still some parts of the world where there is no ground control and such determination may be the only available method. This applies particularly where reconnaissance photogeological maps are being prepared for an area subsequently to be covered by ground survey. In most cases, however, there will be a sufficient number of accurately surveyed points visible on the aerial photographs that can be used as a triangulation network from which the photographs can be scaled. Failing this a first priority of the ground survey should be to set up control points for this purpose. The aerial photographs, line overlaps and mosaics can then be scaled with reasonable accuracy, bearing in mind scale variation with altitude and distortion and changing scale towards the corners of the photographs. However, there will be occasions when either such topographical information is not available or it is inconvenient to undertake a preliminary topographical survey. The scale must then be determined from the photographs alone. The method is simple (Fig. 16), but it depends on information of the flying height and the camera focal length being printed on the margins, and not all photographs have this. Difficulties which then remain are:

1. The determination of flying height above sea level is not always reliable, because it is dependent on knowledge of the local barometric pressure with altimeter correction for this. This information is unlikely to be available in remote areas.

2. The height of the ground above sea level must be known, and there are many regions not covered by accurate topographical survey. Moreover altitude may vary from one part of a region to another. A suitable example can be cited from SW Arabia (chapter 17), where the plains of Lawdar at about 800 m are separated by a steep escarpment from the Mukayras (Audhali) Plateau at about 2000 m (Fig. 16). The photograph 'runs' therefore have two principal scales. Furthermore the aircraft base was at sea level several hundred miles away and there were no meteorological stations to supply information about local barometric conditions.

Height

During reconnaissance mapping it is frequently necessary to determine heights from vertical air photographs, such as those of prominent peaks relative to adjacent valleys, or sea cliffs and other precipices. This permits a much better appreciation of the steepness of the terrain

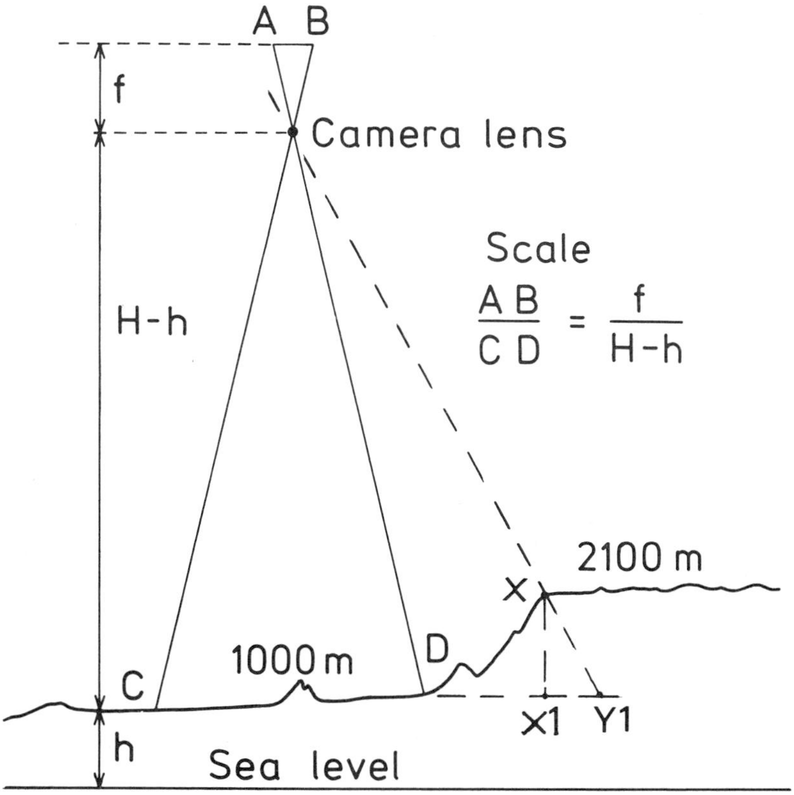

Fig. 16 An aircraft at an altitude H above sea level is flying above ground of variable height (1000 to 2000 m). The photograph scale is given by the simple formula indicated of which f (focal length) and H should be printed on the photograph margin. Problems are: (1) the necessary information is not always printed on the photographs; (2) the height of the ground above sea level may not be known; (3) if one line overlap passes from lower to higher ground there will be scale increase on the higher ground; (4) the position of X on the photograph will be at a distance D–Y1 from D when on a map it would be D–X1.

and of the geological structures, with more reliable estimates of dips. It is generally the case that stereo aerial photographs give a much exaggerated impression of relief, and thus of the angle of dip, and some information on even a few heights can be extremely useful. Descriptions of the methods to be used can be found in books specializing in photogeology (e.g. Allum 1966) but a brief account is appropriate here. The method is simple and is based on parallax difference: the apparent movement of points on the ground when viewed from different positions. For example, on Figs 18 and 19 it is the apparent movement, parallel to the flight line, of the top of the object (T) in relation to its base (B).

Preparation for heighting involves baselining the photographs and recording the mean baseline measurement, which represents the distance travelled by the aircraft between successive exposures (Fig. 14). It is done by joining the centre points of successive photographs. Should the centre point of either photograph fall on some indeterminate area such as the sea or cloud, then it can be plotted as shown on Fig. 17. Two clear topographical features common to both photographs are selected, and arcs a1, b1, a2, b2, enable the centre point of photograph 1 to be located on photograph 2. Most photographs are perfectly straightforward but minor flying and other faults can result in tilt or height variation. If the faults are small they can be averaged, but if large then more refined methods of correction, beyond the scope of this book, have to be used.

The next stage is the measurement of parallax difference. Where altitude variations are small, measurements of parallax difference are small and have to be determined by using instruments fitted with micrometer settings. The stereometer (parallax bar) is perhaps the best known and for its use, which is routine, the reader is recommended to consult Allum's book (1966, p. 94). However, where there are high parallax values (altitude ranges exceeding say 200 m) reasonably accurate values can be obtained by graphical construction without instruments. It may be necessary to do this in the field. Fig. 18 is an illustration of a horse monument as seen on different photographs. The parallax difference can be seen to be (a + b). On Fig. 19 it is (a + b) or (a − b) depending on the position of the tower. Fig. 20 is an illustration from the Kuria Muria Isles, Oman. The method of measuring the height of cliffs in horizontal Tertiary limestone is shown. In geological situations, unlike the tower, the high point will not be vertically above the low point.

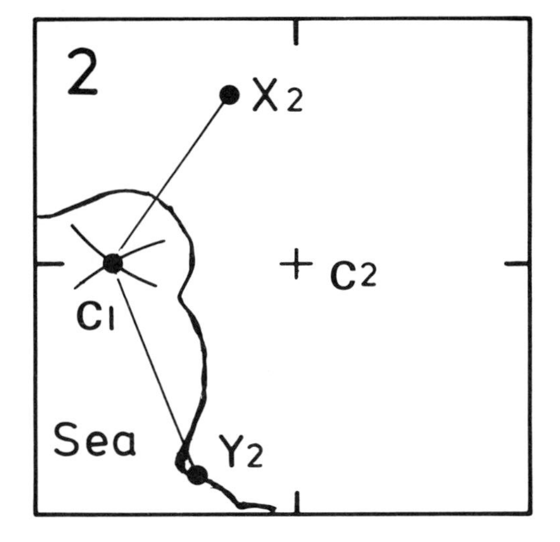

Fig. 17 A method of plotting the centre point C1 on to photograph 2 when it falls on an indeterminate area, in this case the sea. Points X and Y are distinctive topographic features visible on both photographs 1 and 2. On photograph 2 arcs are plotted back to locate C1.

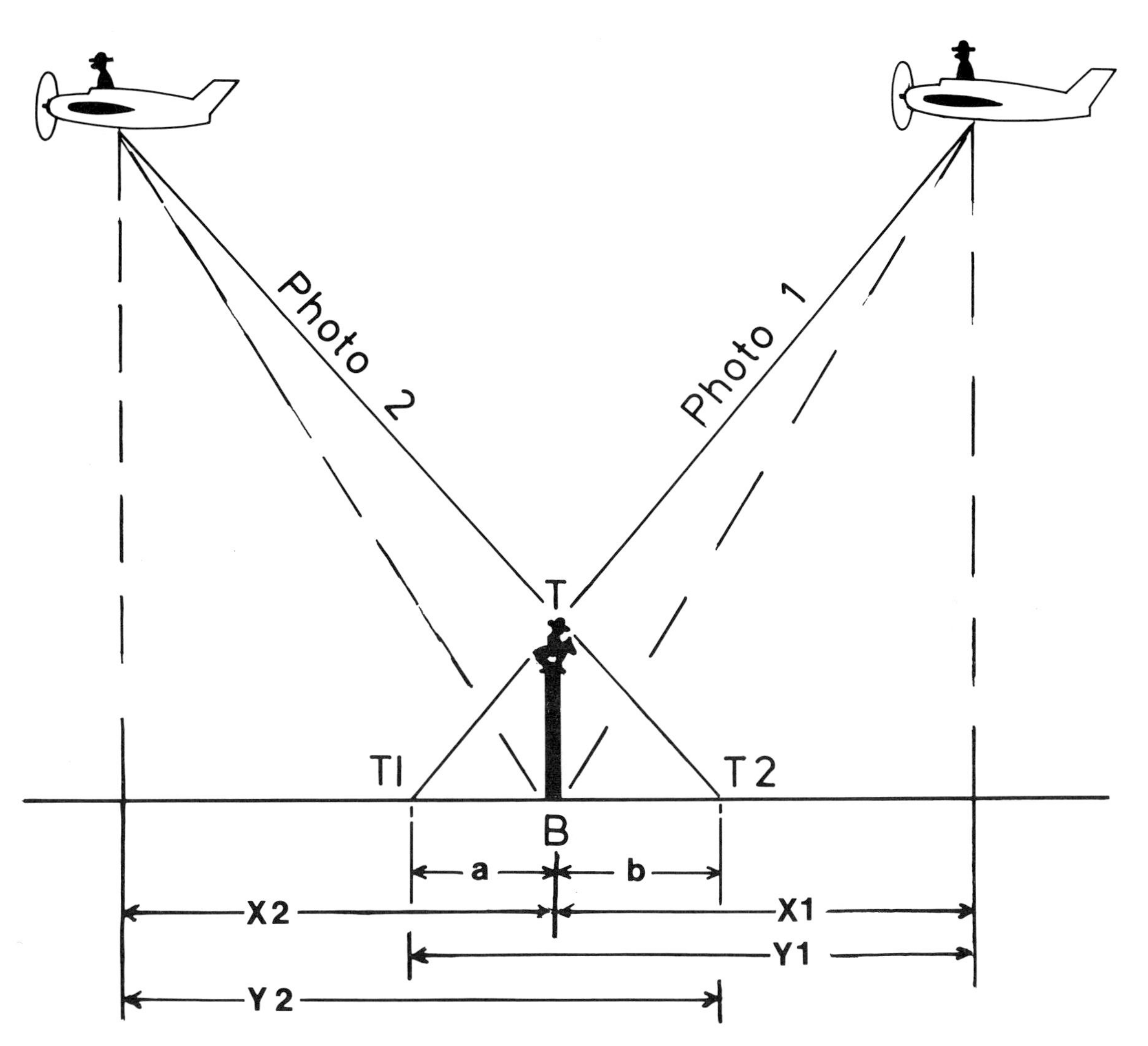

Fig. 18 Parallax and parallax difference. Parallax is the apparent difference in position of a point seen on two adjacent photographs of a line overlap. Thus the parallax of the base of the monument (B) is X1 + X2 (measured from the centre points (C)), and that of the top of the monument (T), which appears on the photographs at T1 and T2, is Y1 + Y2. The parallax difference between the top and the base of the tower is therefore (Y1 + Y2) − (X1 + X2) = a + b. The parallax difference forms a basis for determining the height of objects.

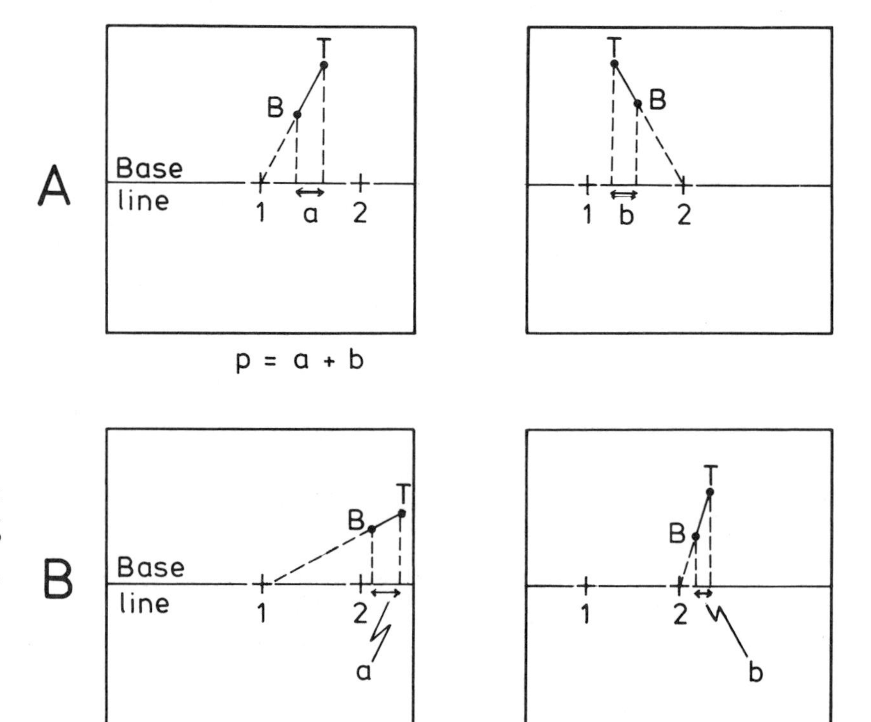

Fig. 19 Consecutive photographs from two line overlaps **A** and **B**, showing determination of parallax difference (compare with Fig. 18). In **A**, a tower with base B and top T is shown, with B – T (vertical) appearing on radial lines from the centre points 1 and 2. The parallax difference between the base and top of the tower is a + b. In **B**, the tower is to the right of the centre point of photograph 2, and the parallax difference is a – b.

Fig. 20 (opposite) Stereopair (enhanced photocopy, see Figs 13, 48 and 116) illustrating ▷ a method for determining the height of limestone cliffs on Hallaniya, Oman, by measurement of parallax difference (see Figs 18 and 19). H, Flying height given on the photograph; P, parallax difference between the cliff top and a ledge at sea level; d, airbase (Fig. 14); h, height of the cliffs.

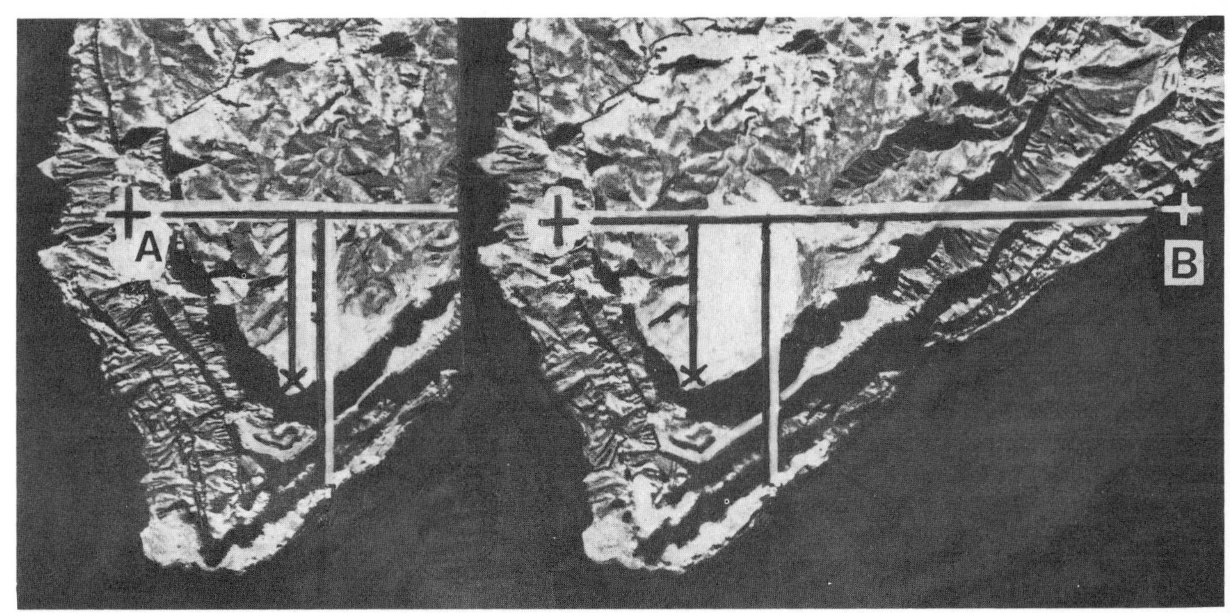

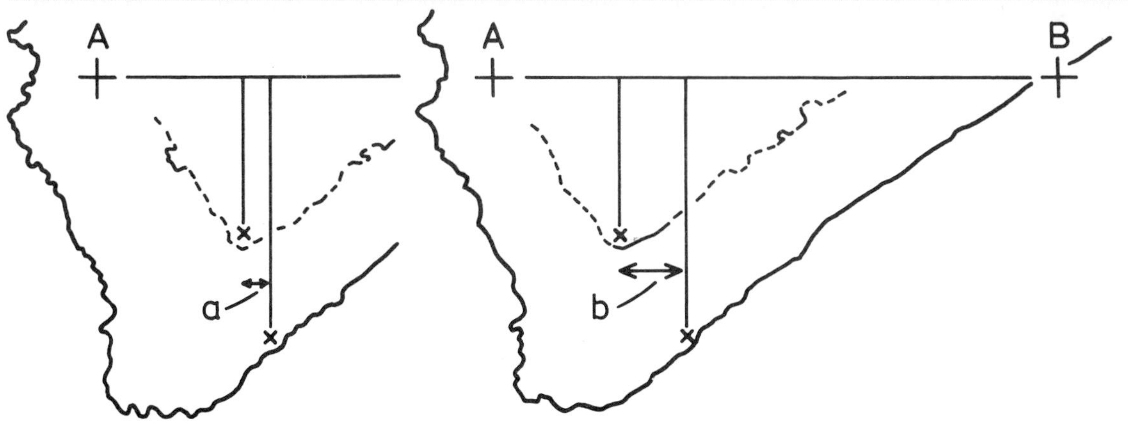

$H = 20\,000$ feet

d :- in mm

$p = b - a$ in mm

$$p/d = \frac{h}{H - h}$$

$$h \sim \frac{pH}{d} \quad \because ph \text{ small}$$

ie $h \sim 1400$ feet

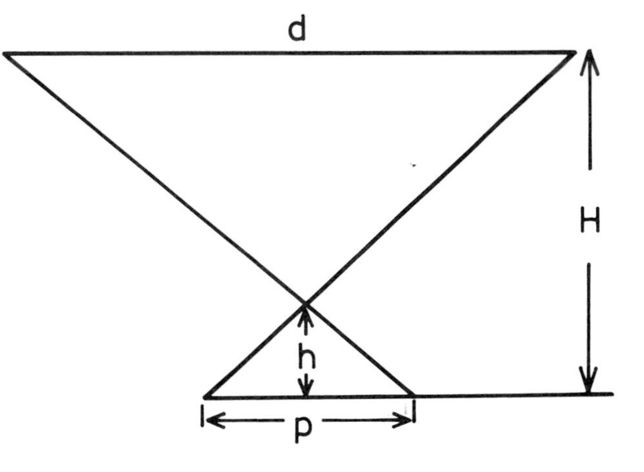

6 Details of field maps

Eventually field data have to be recorded on an accurate map, either a published topographical map or a map made from the aerial photographs if the former does not exist. Examples of different types of field maps are given in Figs 21, 22, 50, 83 and 91 and from these it will be seen that field maps vary in style according to the region, the lithologies and the scale and the detail of the mapping. In some cases an active map will be carried and constructed in the field; in other cases aerial photographs will be used in the field and the 'field map' will be a document compiled each evening at base camp by transferring information from photographs. It is important to be flexible in these matters.

Field data and observations recorded on aerial photographs and maps

In addition to confirmation and correction of the photogeological maps, there will be much data not readily obtained from aerial photographs, but easily obtained in the field. The exact localities where observations are made should be plotted and their details recorded both on the maps and in field notes (Figs 22, 23, 25–27, 50, 55, etc.). It is necessary to emphasize the former since field maps on which a series of locality numbers only are plotted, with explanations left entirely to field notes, can be extremely difficult to follow because cross-reference between map and notes must continually be made (compare Fig. 24 with Figs 22 and 23). A field map should be self-explanatory. Symbols used on maps will vary according to the type of mapping, and once again it is important to be flexible. There is nothing worse than trying to fit one set of conventions to all purposes; the symbols convenient for structural mapping (Figs 12, 22 and 141) have little application to a region of volcanic rocks or Quaternary deposits. The lithologies of exposed rocks should obviously be recorded and lithological (and structural) boundaries plotted. It is often an advantage to shade-in exposed rock with crayon; if the lithologies are sufficiently distinctive a different colour can be used for each. Unexposed ground should be identified if the survey is to be complete; it may consist of superficial deposits such as alluvium, peat or till, and an auger may be necessary to

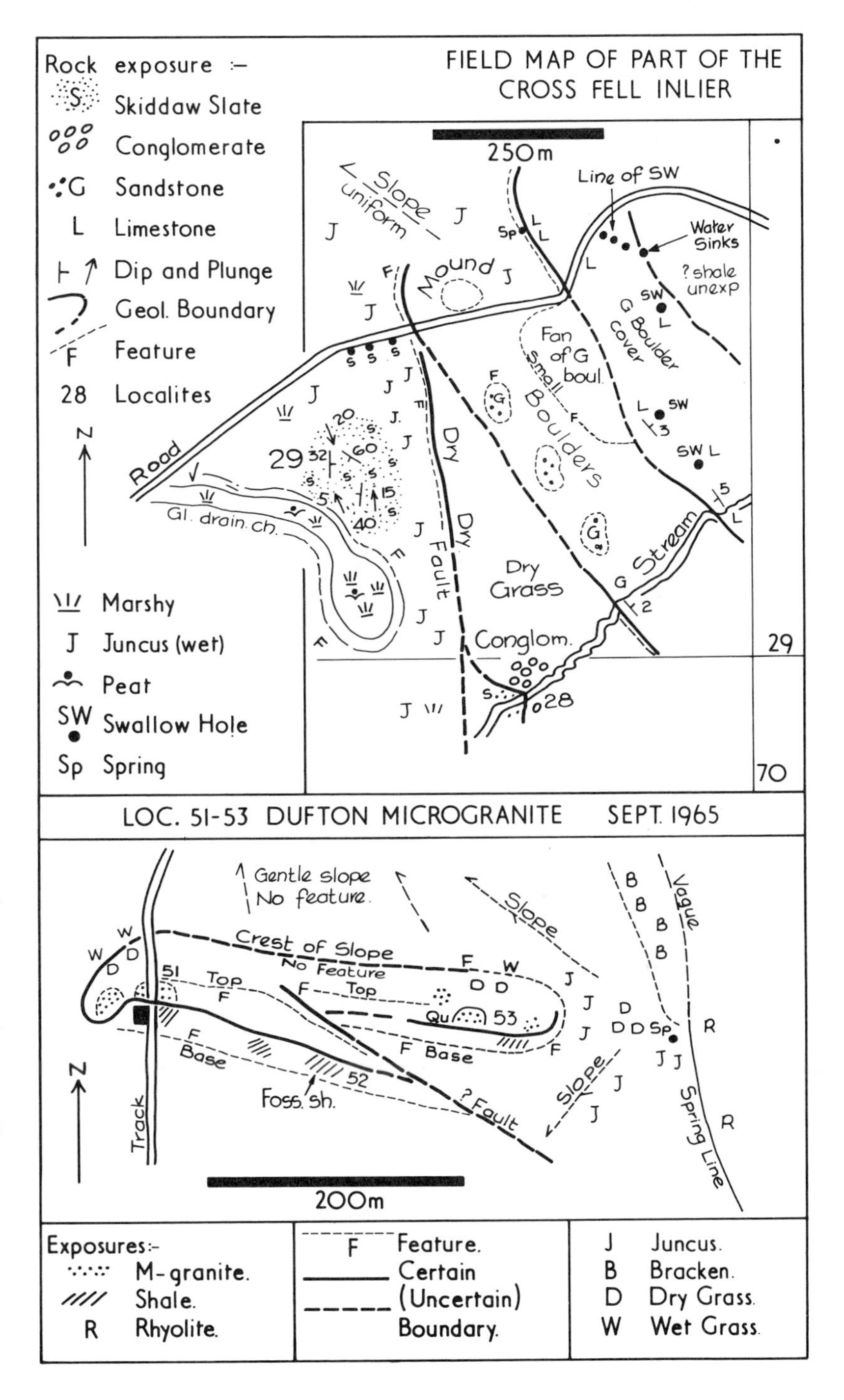

Fig. 21 Copies of parts of field maps of the Cross Fell Inlier, northern England. The upper map shows Ordovician and Carboniferous rocks mapped by a combination of exposure and feature across rather poorly exposed ground. Use has to be made of the nature of the soil, the vegetation, etc. Features are used for drawing in otherwise unexposed boundaries. The lower map shows the importance on occasion of recording the tops and bottoms of features (positive and negative features). Note that in this case shale occupies the lower part of the feature slope and microgranite the upper part. Vegetation also helps in determining the limit of the microgranite outcrop. This map was produced during a Birmingham University mapping class.

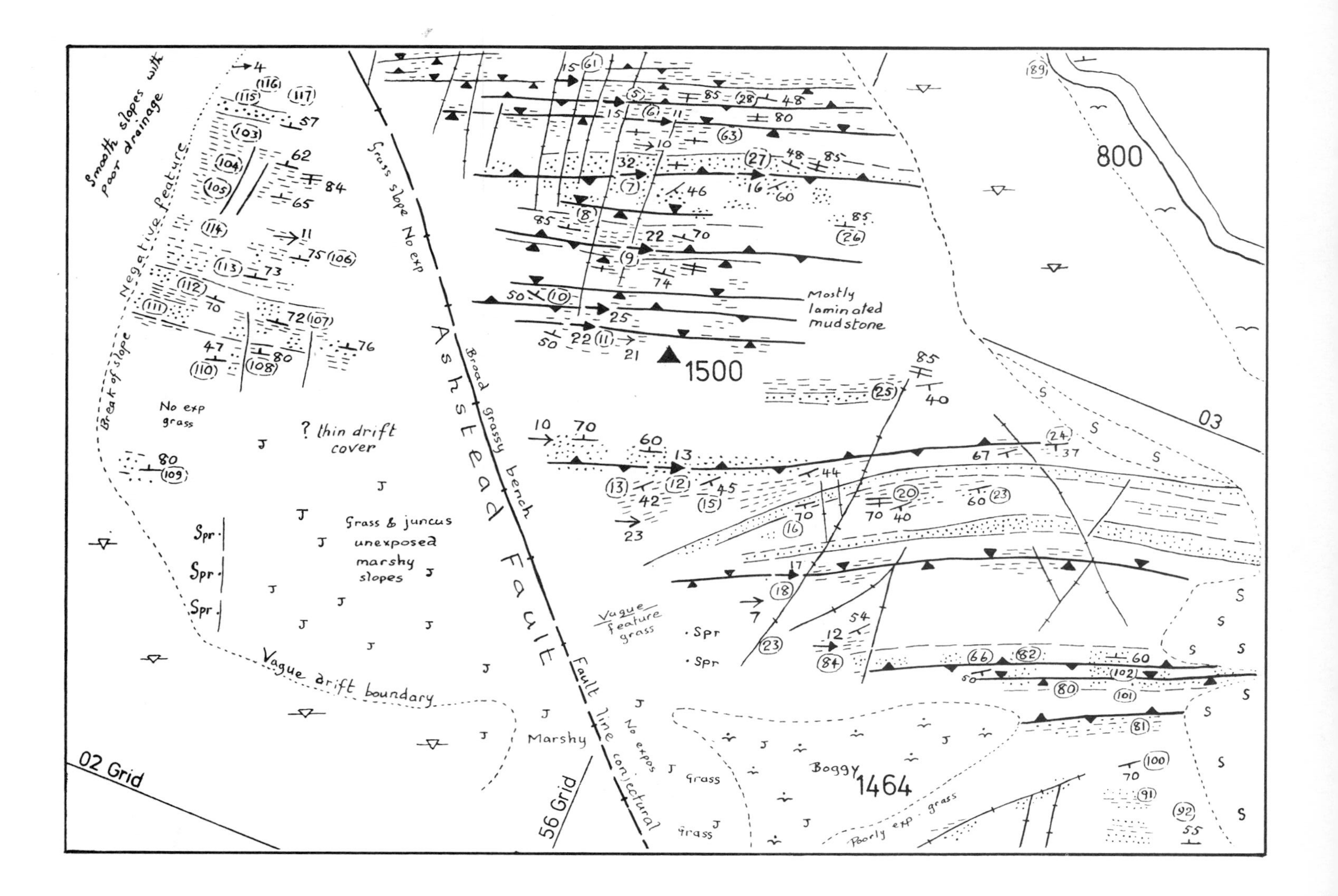

Smooth slopes with poor drainage
Negative feature
Break of slope
No exp grass
? thin drift cover
Grass & juncus unexposed marshy slopes
Vague drift boundary
Grass slope No exp
Broad grassy bench
Ashstead Fault
Fault line conjectural
Mostly laminated mudstone
1500
800
03
Vague feature grass
Marshy
No expos
Grass
Boggy
1464
Poorly exp. grass
02 Grid
56 Grid

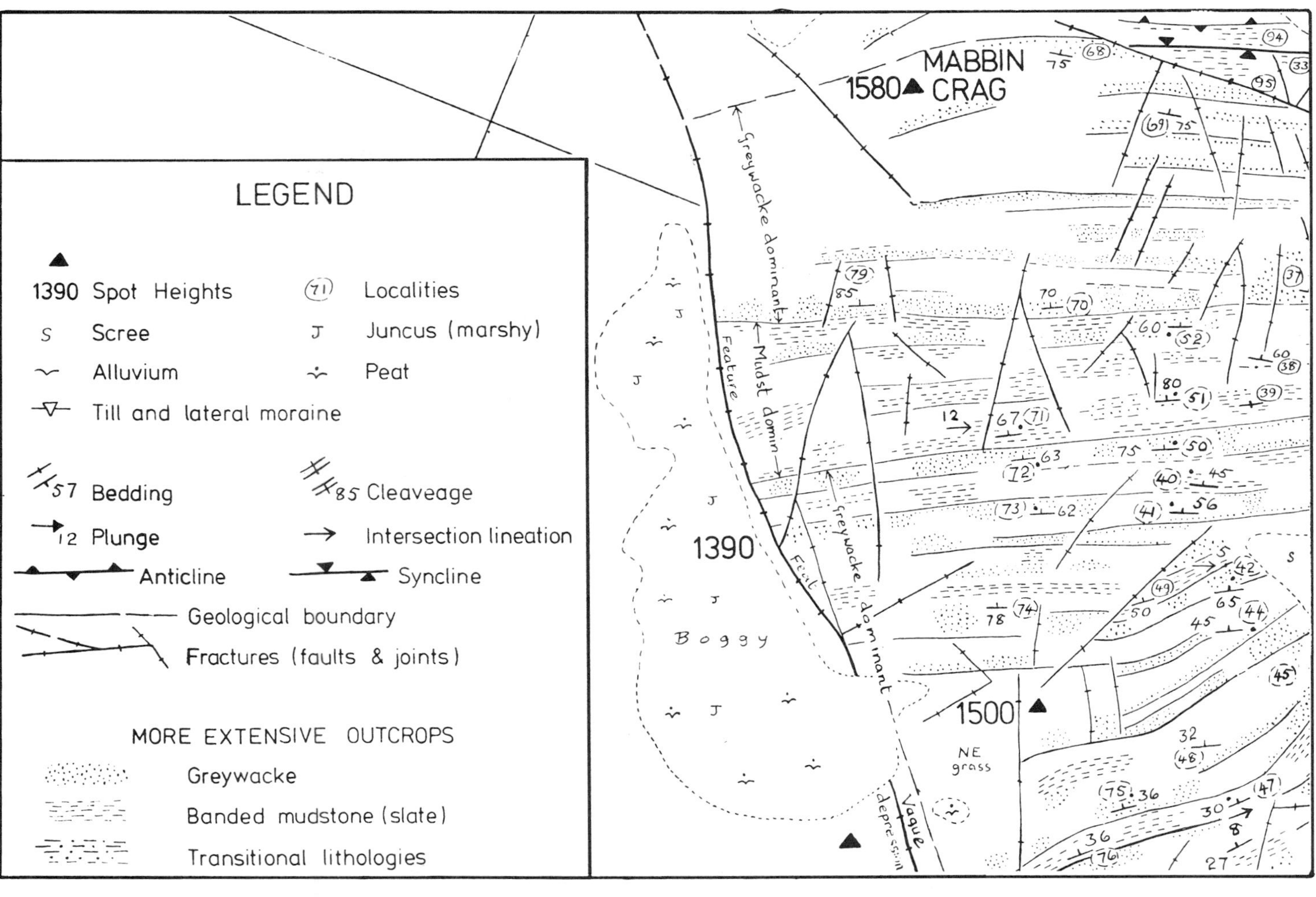

Fig. 22 Copy of part of a field map of the Bannisdale Slates and Coniston Grits (Silurian) near Shap, Cumbria, northern England (Moseley 1968a). The actual field map would make use of different colours of ink for different geological features, and different colours of crayon for the various lithologies. It is not usually possible to record every observation on a field map, and Fig. 23 is a sample page of field notes about this area giving this additional information. Map scale 1:8300. North indicated by the 56 grid line.

Apr. 3.65 Ashstead Fell, Shap.

Local	Bed	Cleav	B/C Int	Mst Joint	Other Joints	Remarks { Grid Ref. 56·03 Readings °T
5 a (557-031)	40.141 32.008			70.260		Plunge 15.072 } AxP 83.344 } Antic in Bd. Mudst. (see also loc 62) Rounded hinge (a) ←8 yds→
b	46.356 55.158	85.171	5.082			Plunge 11.076 Ax Pl. 85.345 30x ENE of (a) in same fold (see also loc 62)
6 a	50.158	80.347	13.075	90.280	See also loc 62	Crags end is Mst. Jt. – 70.268
b	66.360	85.168	20.080			Sync in Bd. Mudst.
c. (557-031)	72.349	85.168	2.078			Plunge 11.075 Ax.Pl. 80.164 (a) (b) (c) NW SE ←10 yds→ ColPhot 2.95 B.W. Stereo-phots 10 (8 & 9)
7 a	90.344			70.257		6 to 7 ~ 70x perp. to strike
b (557-030)	46.126	73.163		70.262		Assym antic in bedded massive greywacke Plunge 32.074 Ax. Pl. 70.153 Col. Phot 2.96 B & W stereo-phots 10 (10.11) NW SE
8 (557-029)	82.355 85.350	73.163	16.079	70.260 *		* Mst jts mark crag line from 7- to 9 — 7. unexposed Sync 8 9 ←~100x→ perp. to strike
9 a	70.004	82.350	50.070			Plunge a-b-22.085 [Bd. Mudst]
b	64.164					Plunge b-c-17.083
c (558-029)	74.358	90.350	25.080			Ax. Pl. b-c-86.172 Bl. W Phot 10 (12) a b c ←1 ft.→

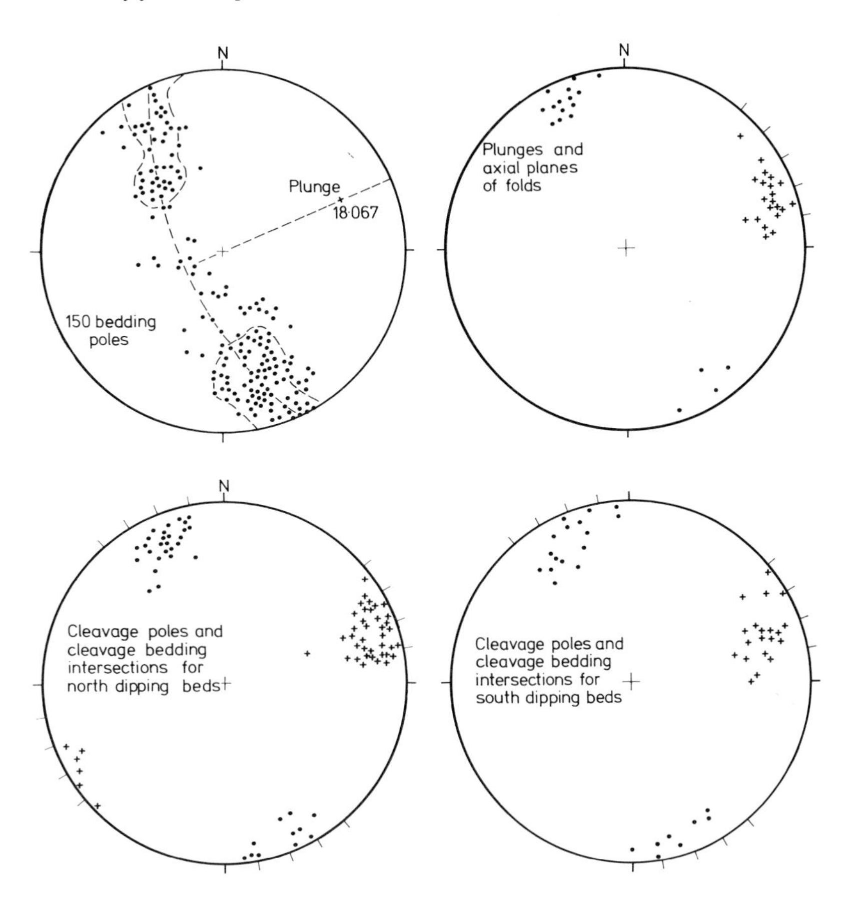

Fig. 24 Plotted data for the area referred to in Figs 22 and 23. When structures are to be analysed, all the information from field maps and notes needs to be plotted stereographically (equal area nets have been used in this case).

determine its nature. Even if details of the drift are unimportant to the survey, or if the ground is drift-free, comments on areas of non-exposure are necessary; for example, the nature of the ground (dry or marshy), the type of vegetation (bracken, *Juncus*) or the content of the soil can all help in subsequent interpretation of the geology. Negative comments such as 'no exposure' should be accompanied by positive observations; there is always something to see (Fig. 21).

Large-scale field maps and plans

Large scale is a relative term but it may be regarded as a scale appreciably larger than that of the base maps or aerial photographs on which the whole area is being mapped. In geologically well-known countries such as Britain, this means scales larger than 1:10 000 (usually about 1:2500), but if reconnaissance mapping is in progress in remote terrain, then 1:25 000 may almost be considered 'large scale'.

◁ *Fig. 23* A sample page of field notes from the map of Fig. 22. The field notes provide the extra detail which cannot be recorded on the map for reasons of space.

Large-scape maps of a few relatively small areas that are critical to the understanding of the geology will be required for parts of practically all mapping projects. During a basic survey to 1:10 000 there will be many cases where larger scales are required:

1. Well-exposed stream sections may facilitate measurement of stratigraphical successions. These may include a variety of lithologies and faunal horizons (Figs 25 and 26). Similar sections may give continuous exposures of complex structures.

2. Quarries and cuttings may yield detailed information similar to (1) above. Sections will be provided by the quarry walls.

3. Natural exposures such as sea cliffs, wave-cut platforms and steep mountainsides may likewise reveal critical relationships that require detailed examination.

Fig. 25 Namurian outcrops near Settle, West Yorkshire, NW England. Sketch enlargement of part of a 6-inch map to facilitate the plotting of additional data for a well-exposed and important area. The sketch map forms part of the field notes (Moseley 1956).

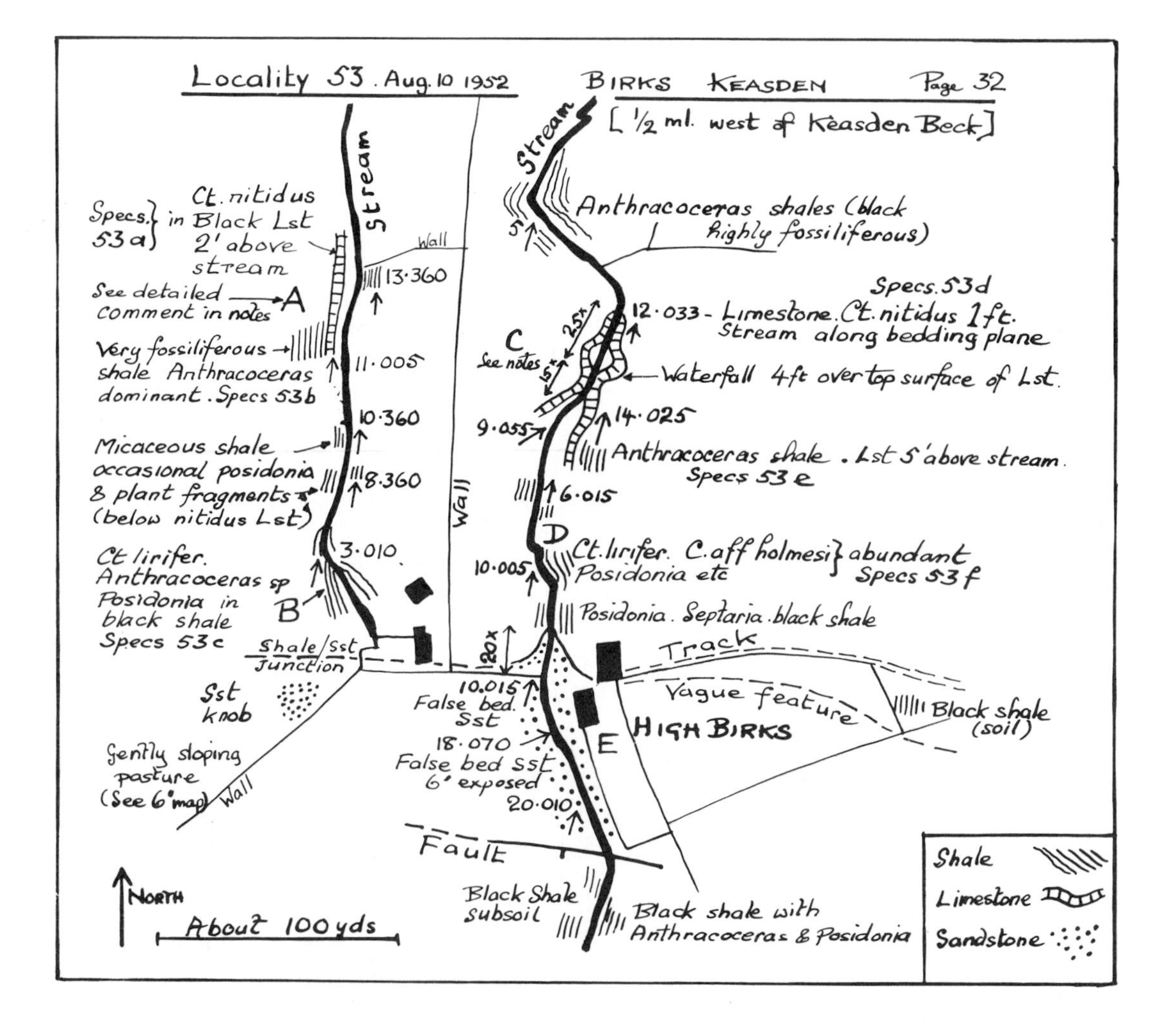

In order to plot information of this kind a large-scale base map is necessary; this can be obtained in several ways depending on the facilities available and the sort of information required.

1. Large-scale topographical maps (say 1:2500) may be available for purchase.
2. An accurate map can be constructed by topographical survey. The most useful methods are plane table (for quarries, etc.) and chain survey (usually tape and compass is sufficiently accurate for geological purposes). A level may also be required (for Quaternary terraces). It must be remembered that field geologists usually work alone and may have to improvise when undertaking topographical surveys. This is not the place, however, to detail the methods of topographical survey, since there are a number of other books that deal adequately with the subject (see Compton 1962).
3. If less accuracy is acceptable it is possible to make approximate but much more rapid enlargements of maps. A square grid is drawn on the smaller scale map, the squares enlarged to any required scale on a blank sheet of paper, the topographical detail inserted by eye, and the geological observations can then be plotted. The same method can be used to enlarge small parts of an aerial photograph. A square grid drawn on the photograph is enlarged (as above) and enough photographic detail transferred to permit the recording of geological observations in the correct relative positions (Figs 27, 81 and 91). The detail may include the bends in a stream course, the outline of a crag or the positions of easily recognized trees. Such maps would frequently be incorporated as pages in field notes and indeed there are many occasions when a series of such sketch maps is the best way of making field notes. An advantage of this method is that it does not require foreknowledge of the areas that are likely to require large-scale maps.
4. If the areas where large-scale maps are needed are known beforehand, photographic enlargements of maps or aerial photographs may be preferred.
5. Ground photography will play an important part in large-scale mapping. Quarries, sea cliffs and other exposures can be recorded; in many cases it is valuable to take stereophotographs as indicated in chapters 8, 15 and 18.

There are many occasions when large-scale plans are necessary; for example, when plotting more details of areas already mapped (Fig. 114) or when investigating details of previously published maps, perhaps with an important principle in mind (Soper 1970).

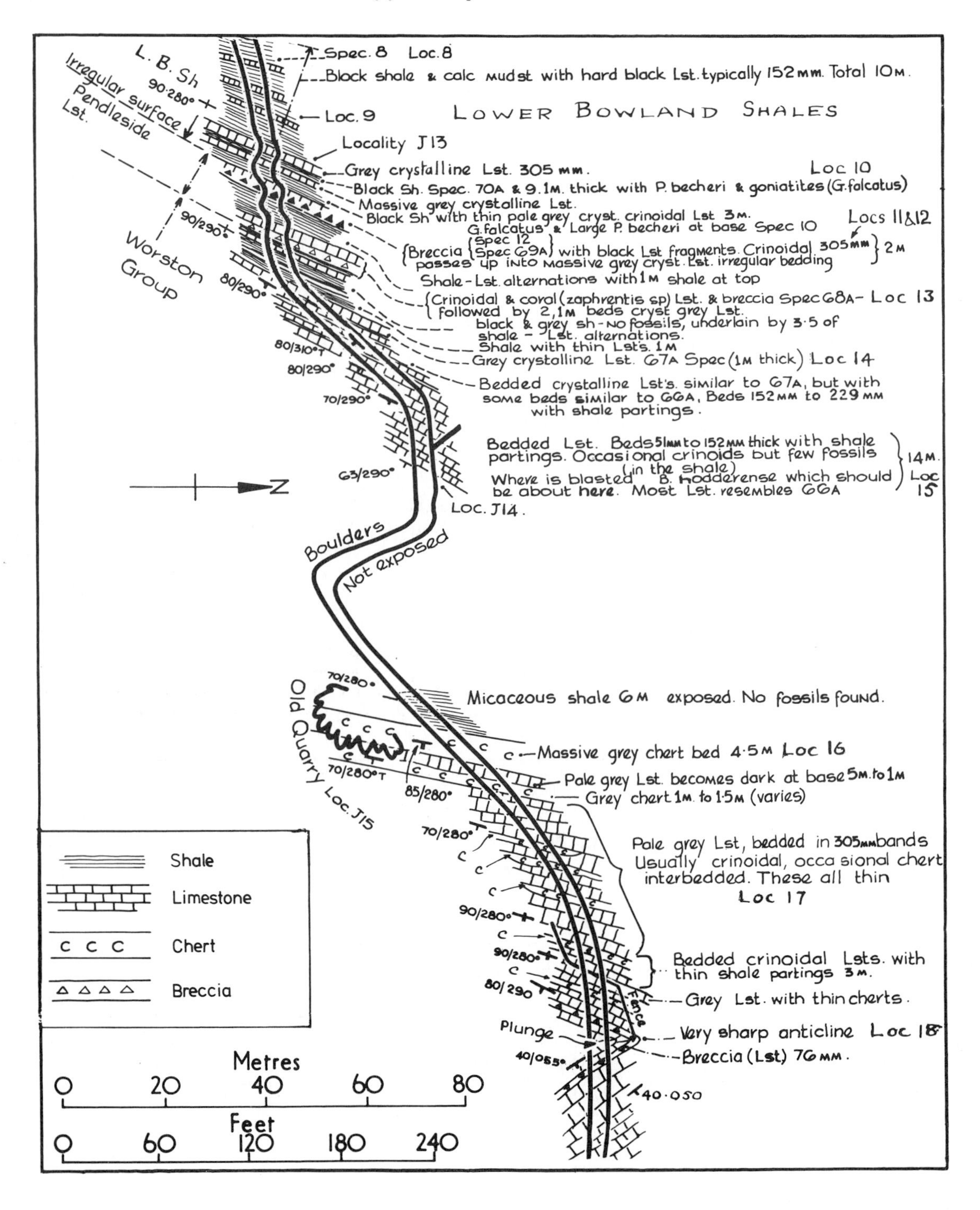

L. B. Sh
Irregular surface
Pendleside Lst.
Worston Group
Spec. 8 Loc. 8
Black shale & calc mudst with hard black Lst. typically 152 mm. Total 10 m.
Loc. 9
LOWER BOWLAND SHALES
Locality J13
Grey crystalline Lst. 305 mm.
Loc 10
Black Sh. Spec. 70A & 9.1 m. thick with P. becheri & goniatites (G. falcatus)
Massive grey crystalline Lst.
Black Sh with thin pale grey cryst. crinoidal Lst 3 m.
G. falcatus & Large P. becheri at base Spec 10
Locs 11 & 12
Breccia {Spec 12 / Spec 69A} with black Lst fragments. Crinoidal 305 mm } 2 m
passes up into massive grey cryst. Lst. irregular bedding
Shale - Lst. alternations with 1 m shale at top
Crinoidal & coral (zaphrentis sp) Lst. & breccia Spec 68A - Loc 13
followed by 2.1 m beds cryst grey Lst.
black & grey sh - no fossils, underlain by 3·5 of shale - Lst. alternations.
Shale with thin Lst's. 1 m
Grey crystalline Lst. 67A Spec (1 m thick) Loc 14
Bedded crystalline Lst's. similar to 67A, but with some beds similar to 66A, Beds 152 mm to 229 mm with shale partings.
Bedded Lst. Beds 51 mm to 152 mm thick with shale partings. Occasional crinoids but few fossils (in the shale)
Where is blasted B. hodderense which should be about here. Most Lst. resembles 66A
14 m. Loc 15
Loc. J14.
Boulders
Not exposed
Micaceous shale 6 m exposed. No fossils found.
Old Quarry Loc. J15
Massive grey chert bed 4·5 m Loc 16
Pale grey Lst. becomes dark at base 5 m. to 1 m
Grey chert 1 m. to 1·5 m (varies)
Pale grey Lst, bedded in 305 mm bands Usually crinoidal, occasional chert interbedded. These all thin
Loc 17
Bedded crinoidal Lsts. with thin shale partings 3 m.
Grey Lst. with thin cherts.
Very sharp anticline Loc 18
Breccia (Lst) 76 mm.
Fence
Plunge
90·280°
90/290°
80/290°
80/310°
80/290°
70/290°
63/290°
70/280°
70/280°
85/280°
70/280°
90/280°
90/280°
80/290
40/055°
40·050
N
Shale
Limestone
Chert
Breccia
Metres
0 20 40 60 80
Feet
0 60 120 180 240

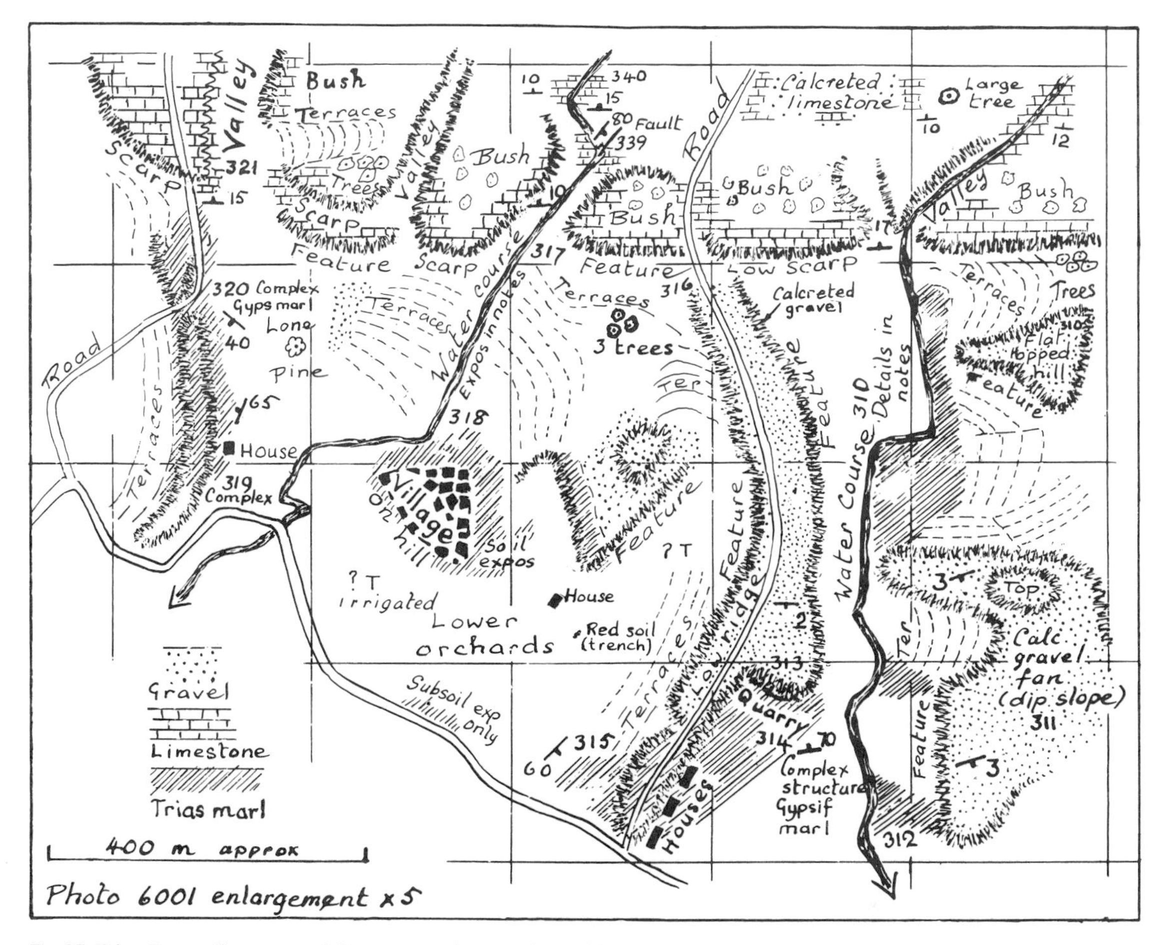

Fig. 27 Trias, Eocene limestone and Quaternary calcrete and gravel in the Pre-Betic Cordilleras of SE Spain (Moseley 1973). Sketch enlargement of part of an aerial photograph to facilitate the plotting of additional detail. In this case the photographic detail on the sketch includes patterns of cultivated terraces, escarpments, houses, prominent trees, etc. The geological outcrops can then be plotted in correct relative positions, in practice using coloured crayons for different lithologies rather than black and white ornament. A square grid is drawn on the photograph using wax pencil, and the squares are enlarged to the desired scale on the diagram.

◁ *Fig. 26* Copy of a section in Lower Carboniferous rocks exposed in Rams Clough, Bowland, NW England (Moseley 1962). This section was compiled in the first place by a tape and compass survey, drawn to scale, and then the details of individual beds were plotted in. The accuracy and time spent on a section such as this will obviously depend on its importance to the overall geology of an area. Where faunal collections are important several weeks may be required. Additional details to those recorded on the section would be found under locality numbers in field notes. For a section in structurally more complex rocks (the Skiddaw Slates of the Caldew, Cumbria) refer to Roberts (1971).

7 Field notes and diagrams

Initial notes taken in the field

Traditionally field notes are written in a reporter's-type notebook, preferably hard-backed and sufficiently small to fit into a pocket. This method is desirable only when either the surveyor requires extreme mobility and cannot carry rucksacks and map cases dangling from the neck or when routine mapping is not the objective. A survey of a difficult mountain climb might come under this heading. On most other occasions it is better to incorporate field notes into a map case as separate sheets occupying one of the pages as shown in Fig. 1. The advantages of such a method are:

1. All the field maps, photographs and notes are contained within a single map case.
2. Notes can be made on large sheets of paper (30 × 20 cm) rather than on small pages in a notebook. This is a great advantage not only for recording data in a systematic tabular form (Figs 23 and 31), but also when sketches have to be drawn and enlarged plans constructed (Figs 25, 26, 28, 30, 55, 81 and 91).
3. Only those field notes relevant to the day's work need be carried into the field, and this minimizes difficulties created by loss. Many a geologist has lost a notebook containing months of work.
4. If the weather is inclement, and notes are made with difficulty, perhaps on wet paper, it is an easy matter with a loose-leaf system to rewrite the day's notes during the evening. Similarly, if bad mistakes are made a page is readily scrapped and a new one started.
5. If exposures have to be revisited at a later date and further details recorded, new pages can be inserted in the correct position. It is an advantage to have all the observations for one locality in the same part of the notes.
6. Photographic illustrations can be mounted on separate pages and inserted opposite the exposures to which they refer.

Completed notes

When a field topic has been completed or a specific area has been mapped, then loose-leaf notes relating to it can be gathered together to

Fig. 28 (overleaf) Copy of a field sketch of the Lilloise Massif, East Greenland. The massif is composed of Tertiary basalts intruded by a layered gabbro–peridotite complex (Brown 1973). The side of each square represents the length of a pencil held at arm's length. This method of sketching makes it easy to place features in their correct relative positions, and to obtain more accurate scale drawings. Photographs should accompany a sketch and a series of stereophotographs were taken to cover this panorama using lenses of differing magnification. Fig. 29 shows one of these photographs for comparison. The absolute accuracy of a sketch is not required since this would use too much valuable field time, but the accuracy should be sufficient for localities plotted on the sketch to be located on the photograph. BW and Col refer to black-and-white and colour photographs. The compass orientations of a number of prominent points should also be recorded since this will later assist in the compilation of a map. These are shown as 225°T etc.; M2–8 refer to unnamed peaks. (See also Figs 127 and 128.)

Fig. 29 (overleaf) The Lilloise Massif, East Greenland. The photograph was taken from a locality close to that from which the rough sketch of Fig. 28 was drawn. Note that the sketch is sufficiently accurate for the principle topographical features of Fig. 28 to be related to the photograph. In fact stereophotographs were taken making it possible to locate details not visible on single flat photographs. An example of the value of stereophotographs is indicated by point A. On a single photograph the Lilloise summit ridge appears to be continuous with the ridge to the left of A, whereas the stereophotographs show them to be quite separate with a major valley between A and the Lilloise ridge (very important should one consider climbing the mountain by that route). The near horizontal layers are basalt lava flows, the gullies are near vertical dykes and a small section of the Lilloise mafic intrusion is visible on the left (Brown 1973; and chapter 18).

make a field notebook, the easiest and most obvious method being to staple the sheets together. But it must be stressed that field notes may have to be used at a later date by other workers, and even if the original surveyor comes back to them after two or three years details will have been forgotten. It must therefore be easy to refer to any desired aspect of the geology of the region. This can be done as follows:

1. Each evening notes should be clarified by underlining headings and important items, and by inserting cross-references to other localities where there are similar rocks and structures. This is equivalent to the clarification of field maps each evening by 'inking in' all the data recorded during the day's fieldwork. If this is not done promptly details will most certainly be forgotten and observations so obvious at the time will later appear to be vague and confused. It is also important to number the pages and prepare an index and contents so that any desired item can be referred to with no delay.

2. The pages can then be stapled together to make a permanent notebook, perhaps with a title page, and with an envelope at the back containing relevant field maps. This could be done when the geologist has returned from the field. Aerial photographs and overlays are usually too bulky to be contained in a field notebook and are best stored separately. Similarly many ground photographs, enlarged for laboratory study, will be too large to be incorporated in a field notebook and are best kept in a separate folio. Stereopanoramas made of the Lilloise Massif of East Greenland and of the Akrotiri cliffs, Cyprus (Moseley 1976) both exceeded 2 m in length (chapters 15 and 18).

Tabulation of field notes and annotation of field sketches and photographs

Tabulation

It is important that data recorded in field notes should be easily recovered for analysis at a later time. In order to facilitate this, notes should be arranged in a tabular form (Figs 23 and 31). The actual layout of notes will naturally depend on the type of survey, but in all cases it should be possible to extract any set of readings quickly. For example, it may be necessary to compile a list of cleavage orientations or of lineations, or comparisons may be required between many outcrops of one rock formation visited at different times and therefore scattered throughout the field notes. Rapid recovery of data of this kind is certainly made easier if single columns are reserved for the same type of information, and also if there is a table of contents and index to accompany the notes.

Field sketches and photographs

Field sketches always form an important part of a geological survey. They should be accurate, but it is a mistake to insist on perfection and beauty since not every geologist is a good artist, and time can be wasted in making beautiful field drawings. The obvious compromise is a moderately quick line sketch, with the correct relations between features preserved as far as possible by using the method of 'squares' (Figs 28, 30 and 55). The simplest procedure is to hold a pencil at arm's length, with the pencil length equal to one of the sides of the square, and then to draw within these squares. This allows the topographical and geological features to be recorded in their correct relative positions. This type of construction may seem unnecessary, but it is in fact

Fig. 28

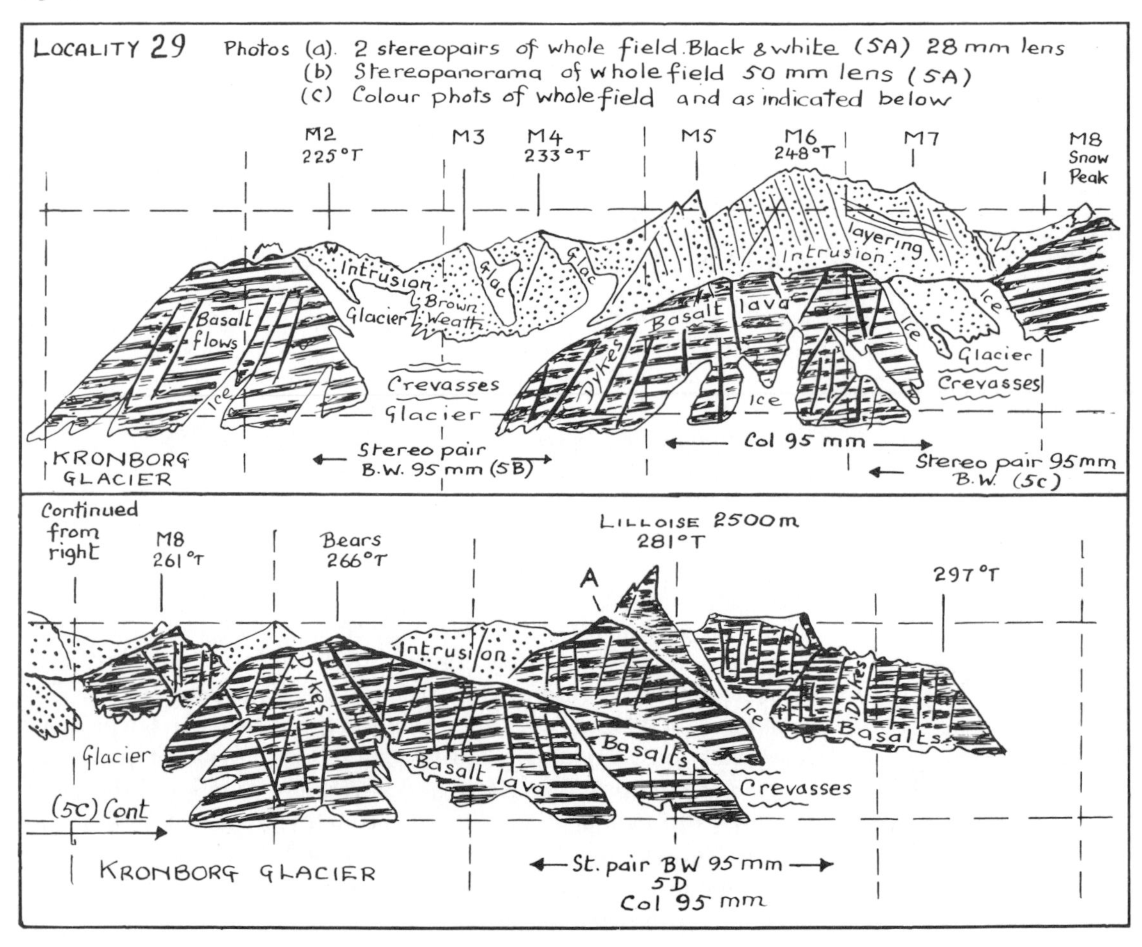

surprisingly difficult to attain the correct proportions in any field sketch. Those who have had the opportunity to compare their field sketches with photographs will readily understand this difficulty, and indeed a true test of any field sketch will be subsequent comparison with a photograph (preferably a stereophotograph) and the identification on the photograph of the features shown on the sketch (compare Fig. 28 with 29). In addition, the sketch should emphasize and record items which cannot be derived from photographs alone, such as thicknesses of strata, lithologies, the exact position where measurements are taken, and the compass orientations, elevations and depressions of prominent landmarks. In this way sketches and photographs will complement each other; both are indispensable.

Fig. 29

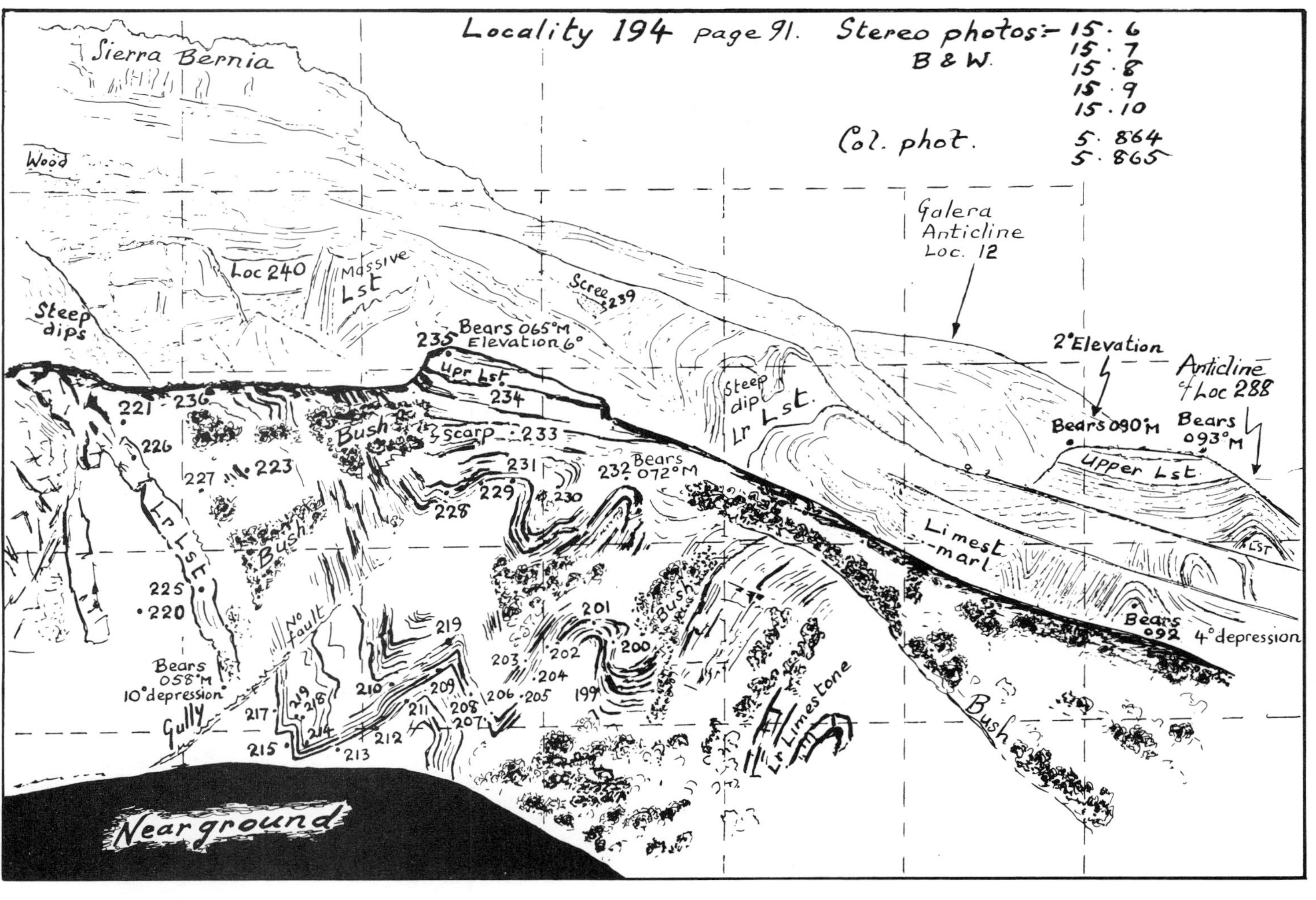

Fig. 30 Field sketch of disharmonic folding in Cretaceous marls and limestones of the Pre-Betic Cordilleras of SE Spain, drawn by the method of squares (see Fig. 28) and corrected from stereophotographs which in this case were already available from an earlier reconnaissance. Note that compass bearings and elevations and depressions of different points are given. The numbers refer to localities subsequently investigated where detailed measurements were taken. These were plotted on the sketch at a later time (see Fig. 31 and Moseley 1973).

Fig. 31 Field notes referring to details of two localities shown on Fig. 30.

Page 98

Aug 7.68 GALERA. cf Loc 194. Air phot 6316–6317 Run 2

Local.		Bedding	Plunge	Ax. Pl	Remarks. [Readings °Mag]
230	a.	35·152	17·085	70·160 [difficult]	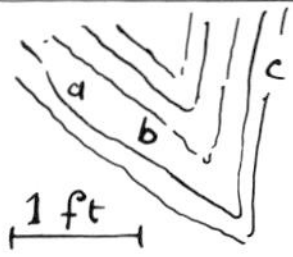 Thin bedded marly Lst. Very sharp syncline
	b.	40·152			
	c.	90·176			
231	a.	90·172	g–h 4·080	Difficult to measure	Lst as 230. 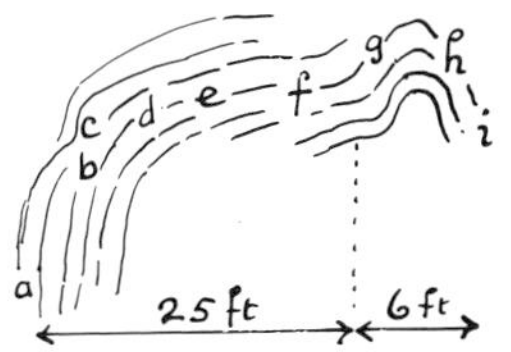 Not possible to phot-ograph
	b.	80·200			
	c.	70·190			
	d.	56·215	Lower		
	e.	30·230			
	f.	34·240	20·080 (unrel-iable)		
	g.	63·004			
	h.	25·110			
	i.	70·170			
232	a	60·021			
	b	85·015			

8 Stereo ground photographs and preparation of sections using photographs

Field sketches and photographs can be combined with measurements of quarry and cliff faces so that it is possible to compile scale sections quickly. This operation starts with field sketches and stereophotographs (below), continues with other field measurements and is completed either at base or in the laboratory at the end of a field season. Views of distant cliffs can be extremely useful for locating the positions of obscure outcrops so that they can be visited in the future (Figs 32 and 33). Close-up photographs of a large area of horizontal surface may also be required and can be taken as shown on Fig. 144 so that a mosaic is obtained.

Stereo ground photographs

Stereo ground photographs can be taken in the field, either as stereo-pairs or as more extensive stereo-line overlaps; the former are useful for small outcrops such as quarry faces, whereas the latter may cover several miles of a cliff (Mitchell *et al.* 1972; Moseley 1972b). The resulting three-dimensional stereo model not only adds appreciably to the perception and interpretation of the geology but, equally important, dramatically speeds up the whole process of survey. It is important to remember that the human eye cannot resolve stereoscopically at distances much greater than 400 m, but the effective 'eye base' can be increased by taking photographs of the same section from adjacent positions. Stereophotographs of this kind will yield more information than can be seen directly by an observer from the same viewpoint.

The methods are simple and basically the same as those used in aerial photography (Allum 1966; Lattman and Ray 1965). They are listed below under headings likely to be of value to field geologists. There are, of course, more specialized procedures using instruments such as the photo-theodolite, with the data programmed for computer so that rapid and accurate computations of distances and the orientation of planar surfaces can be made. These methods are not considered here since they are unlikely to apply to normal field survey.

Line overlaps

Line overlaps may be taken by traversing parallel to a line of cliffs; for example, if structures in sea cliffs are to be recorded the line overlap can be taken from a boat. Generally the percentage overlap needs to be somewhat greater than is the case with vertical aerial photographs (at least 80%) because the 'relief' (the distance between foreground and background) tends to be relatively greater. It is possible to make a detailed record of miles of cliff in quick time using this method (chapter 15).

Stereopanoramas

Stereopanoramas are perhaps more easily practised. Two panoramas are taken of the same cliff line or mountainside, with the camera stations some distance apart (say 100 m if the mountainside is several miles distant), with the line joining the camera stations parallel to the mountainside or cliff (Figs 29, 102, etc.). One of the panoramas will have to be sectioned and folded to facilitate viewing with a stereoscope. Non-stereopanoramas and mosaics are, of course, also useful as can be seen from Figs 69, 71, 102, 110 etc. (it is inconvenient to reproduce stereopanoramas in this book since they would occupy too much space).

Stereopairs

Stereopairs are the same in principle as stereopanoramas. The distance between camera stations for each photograph of the stereopair will depend on: (1) the distance from the camera to the section being photographed; and (2) the depth of the field of view which it is intended to record. Thus distant precipices or mountainsides with little 'relief' (that is, the whole of cliff-line is about the same distance from the camera) may require camera stations 100 m or more apart. On the other hand if there is a considerable depth to the field of view, comparable with high oblique aerial photographs, it is better to have smaller camera spacings; close-up photographs of quarries or road cuttings may require the camera spacing to be as little as 1 or 2 m.

Other methods of section drawing

Two other simple methods of drawing sections can be done in conjunction with original field sketches (page 42). The first is to project colour transparencies on to drawing paper at the scale required and to draw the geological lines direct so producing an accurate sketch such as that shown on Fig. 34 (see also Fig. 95). Lens distortion will, of course, result in some inaccuracy, but the diagram will be a valuable addition to a field sketch. Field sketch data such as stratal thicknesses and other items listed above can be plotted on to the diagram, which can then be taken into the field and additional data can be added. The second method is to photocopy the original photographs of the exposures and to combine these with information from field notes and sketches.

A
B
Fig 33

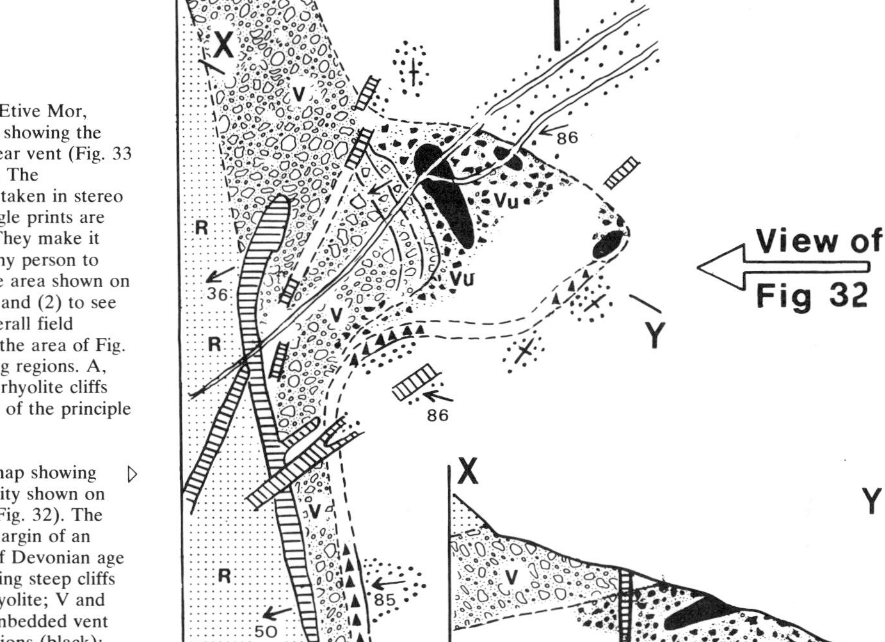

◁ *Fig. 32* Buchaille Etive Mor, Glencoe, Scotland showing the position of the linear vent (Fig. 33 and Hardie 1968). The photographs were taken in stereo (although only single prints are illustrated here). They make it possible: (1) for any person to locate and visit the area shown on Fig. 33 with ease; and (2) to see at a glance the overall field relations between the area of Fig. 33 and surrounding regions. A, Vent breccia with rhyolite cliffs above; B, position of the principle inclusion.

Fig. 33 Outcrop map showing details of the locality shown on the photographs (Fig. 32). The locality is at the margin of an early linear vent of Devonian age with rhyolite forming steep cliffs to the east. R, Rhyolite; V and Vu, bedded and unbedded vent breccia with inclusions (black); triangles, explosion breccia; lines, post-volcanic dykes; dots, quartzite (after Hardie 1968). ▷

Using black and white inks (or a black felt-tip pen and process white with a fine brush), important geological lines can be emphasized (in black), and shadows and other irrelevant detail can be painted white. In this way accurate diagrams can be drawn (see Figs 43, 56, 70, 76, etc.).

The methods outlined in this chapter can be used to compile geological sections quickly. For example, using a motor launch 5 km of the coastal cliffs of Akrotiri, Cyprus, were recorded stereoscopically in 20 minutes (Moseley 1976 and see chapter 15) and it required very little time to record the folded Silurian section alongside the M6 motorway near Tebay in Cumbria, northern England (Mitchell *et al.* 1972). Diagrams were later prepared from these photographs, and information on lithologies and thicknesses was added to them. It is not, of course, always possible to revisit a locality—this was true of the Akrotiri section—but it was possible to revisit the M6 section, and the prepared diagrams and photographs, used in the same manner as field maps and vertical aerial photographs, made subsequent recording of additional data an easy matter.

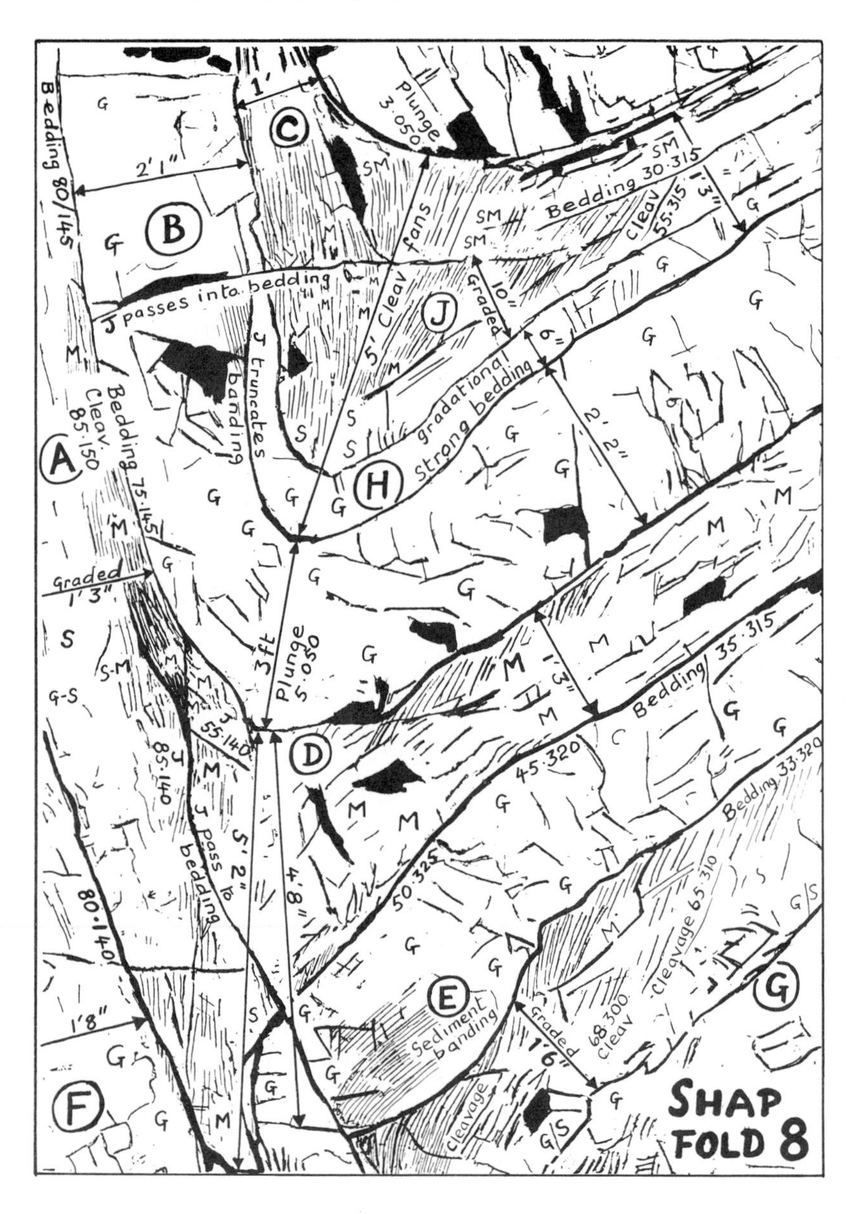

Fig. 34 Diagram of a syncline in Silurian greywacke and mudstone, Shap Fell, NW England, prepared by projecting a colour slide on to drawing paper and tracing details of the structure. The diagram was taken into the field and further details added. Compare with the methods of drawing from stereophotographs (Figs 42 and 44) and by enhancement drawings of photocopies (Fig. 43 and text).

9 Reconnaissance surveys and expeditions

In Part II some of the methods of tackling reconnaissance surveys of comparatively remote regions are described. There is, of course, infinite variety in surveys of this type. The terrain can vary from Arctic mountains to arid deserts and steaming jungles; the resources may be plentiful, as with some government surveys and commercial companies, or they may be meagre, as is the case with many university expeditions. The objectives will also vary since the survey may be concerned with hydrogeology, oil, mineral deposits, fundamental scientific research, or some other purpose. It is impossible to cover the whole range of reconnaissance surveys, but in Part II several contrasting types have been selected as case histories.

Selection of the party and field procedure

It is often the case that an expedition party will live in isolation for weeks, or even months, and it is well known that even the best of friends can become bitter enemies in such situations. People with small irritating habits, those who are thoughtless, dogmatic, and those who easily become exasperated are least likely to foster a happy expedition atmosphere. Optimists are much to be preferred and it goes without saying that all should be enthusiastic, resourceful and prepared for a team venture rather than individual glory.

In remote regions individuals should never travel alone. To cross a crevassed Arctic glacier requires a minimum of three properly equipped persons, and an expedition to such a region should certainly consist of at least six members. Surveys to deserts are equally demanding, and generally require two Land Rover-type vehicles, preferably with trailers, comprehensively stocked with petrol and water. A large selection of spare parts should be taken, and the expedition should include someone with mechanical competence.

If one may be a little cynical it would be to observe that practically all geologists experienced in expeditionary work would support such obvious precautions, but it is a matter of common observation that this can be merely lip service, and many of those same people often take astounding risks.

Example of a reconnaissance survey

Fig. 35 refers to a day's fieldwork during a reconnaissance survey, and illustrates the difference between this approach and a detailed survey. Three people were involved, one staying with the vehicle and two undertaking the reconnaissance traverse, which was across basement metamorphic rocks in desert terrain. From the aerial photographs the rocks were easily identified as varieties of gneiss, outcropping along a fairly steep mountainside some 600 m high. The vehicle was left in the valley at A; the different varieties of rock were then sampled and measurements were taken along a traverse at right angles to the strike. The ridge was followed and another traverse was made at B, where the vehicle was rejoined. The combination of photogeological map, measurements of foliation and other structures, specimens collected and ground photographs resulted in a fairly reliable reconnaissance map (Moseley 1971b).

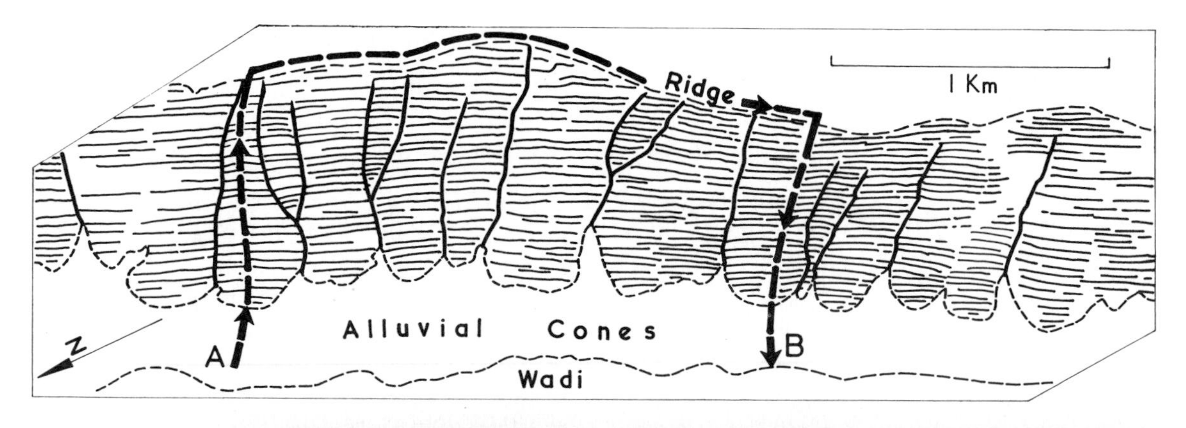

Fig. 35 An example of reconnaissance mapping on basement gneisses near Beihan in South Arabia (Moseley 1971b). Aerial photographs (oblique or vertical) facilitate the selection of a route for a single day's geological traverse. A–B represents such a route. The vehicle was left at A and a traverse made up the mountainside (about 600 m high) during which structural and petrographic observations were made and specimens collected. Similar observations, made while descending from the ridge to the vehicle waiting at B, then allowed the construction of a geological map based on the aerial photographs. (Crown copyright/RAF photograph.)

10 Preparation of a degree thesis

Although there will be many differences in detail, similar principles apply to the preparation of a BSc, MSc or PhD thesis, a report for an employer, or a paper for publication in a geological journal. Most experienced geologists should find the following comments straightforward and obvious, but undergraduates are inexperienced and it is important that they pay a great deal of attention to presentation. It is my opinion that thesis preparation should be regarded as part of the degree training, just as much as a laboratory class, and there should be instruction and supervision relating to it, although the final edition should be entirely the work of the student. Thesis writing will serve as an introduction to future geological writing, whether it be company reports or papers for publication. Methods can be learned in the first place by consulting published papers on topics similar to that of the thesis, but guidance from a supervisor is also necessary. I do not agree with the philosophy that a student should be left entirely to his or her own initiative, either during the work in the field or the writing-up stage.

With these thoughts in mind some important points can be outlined.

Date of submission

It is usually required that a thesis be handed in by a specified date. In my experience perhaps one in five undergraduate theses are handed in late. This suggests unreliability and is bound to be reflected in a supervisor's report to a potential employer. Company reports have deadlines which must be observed.

Length

It is a great temptation to believe that the worth of a thesis is directly proportional to its length. This is not so. It is far more difficult to write concisely and to condense a great volume of information into a few pages so that no important details are omitted. This is a skill which should be acquired, and it needs practice. As a generalization most published papers are short, rarely more than 10 000 words and often much less. An undergraduate thesis should abide by these principles, both in the written text and in the numbers of the text figures and photographs presented. The latter should be carefully selected to illustrate and clarify important features for which the written word is inadequate. A published paper of, say, 8000 words will rarely have more than 10 to 15 text figures and 4 or 5 photographs; admittedly this is usually at the insistence of an editor who has to watch his budget carefully, but, to repeat the point, thesis writing is a preparation for just this sort of publication. The place for numerous explanatory photographs and diagrams is in field notes.

Illustrations

Text figures and photographs are certainly necessary, and if properly selected and drawn will illustrate the geology with greater clarity than most written accounts. There is an art in constructing text figures and geological maps; not all geologists are artists, but it is easy to apply some common rules.

In the first place a thesis should always start with an introductory map showing the location of the area or areas being described, and in most cases it is useful to have a vertical section showing the succession of strata (Fig. 36) and horizontal sections to show details of structure. If possible it is best to draw the latter with the same vertical and horizontal scales, but if the structures are gentle, vertical exaggeration may be necessary. Examination of Fig. 37 will show some of the problems associated with section drawing and scale. Enlargement maps to show details of areas critical to the understanding of the geology may also be necessary (Fig. 38) and diagrams to illustrate hypotheses proposed are often desirable (e.g. Fig. 87).

The use of ornament on maps is illustrated by Fig. 39: no ornament is used on A; standard ornament is used on B, but with no consideration of geological structure; whereas on C, hand-drawn ornament distinguishes the minor folding at Z, the unconformity separating Z from X and the unconformity below Y. Map C reveals these features at a glance, maps A and B do not. It is also important that maps should have a scale and a north point (as in C); A has neither. The maps accompanying most undergraduate and many PhD theses, use different colours (paint or crayon) for different geological formations. The colouring is usually done badly; since most published maps in journals are unlikely to be in colour (because of cost), I suggest that thesis maps should first be drawn in black ink and then ornamented. Hand-drawn ornament is preferable (see Fig. 40), since printed adhesive ornament (Figs 38, 41 and 63) is rather expensive. Colour can always be added as a further improvement for the 'top' copy of the thesis, but if this is done it is advisable to keep to pale shades unless the student is a good artist; whatever is used the maps should be absolutely clear without addition of colour.

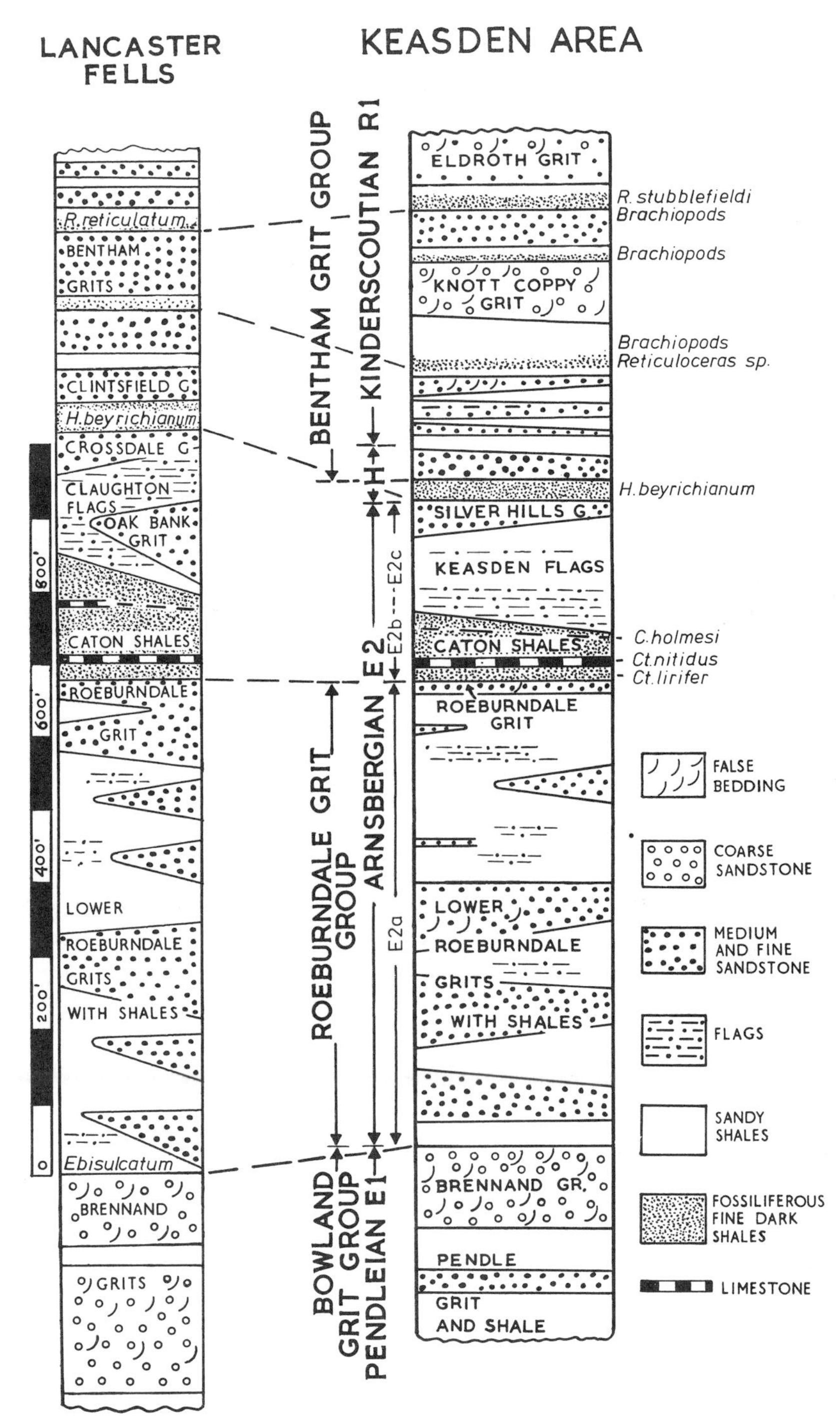

Fig. 36 A vertical section to show the strata exposed within a mapped area of Millstone Grits in West Yorkshire, NW England and its correlation with an adjacent region (Moseley 1956). Such a diagram would form a text figure near the beginning of a report or paper. (See also Figs 25, 52, 54 and 63.)

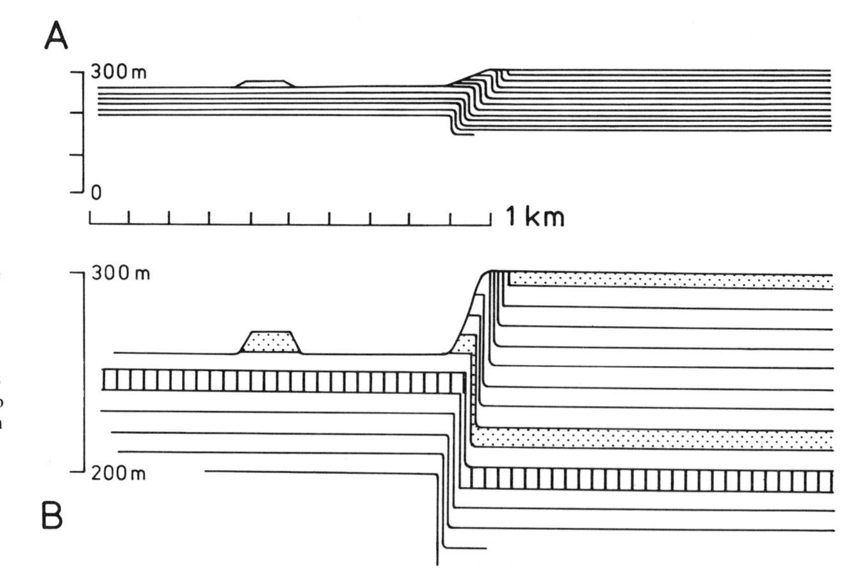

Fig. 37 Problems of section drawing. Two sections across the same monoclinal fold (horizontal and vertical limbs). In **A** the same horizontal and vertical scales are used, showing the true geometry of the structure, but details of individual members of the sequence are not easily seen. In **B** the vertical scale is exaggerated so that details of the stratigraphy can be illustrated, but thicknesses of horizontal strata are measured by the vertical scale, and those of vertical strata by the horizontal scale. The result is an awkward-looking structure.

There is one other long-standing tradition in Britain that thesis maps are drawn on an Ordnance Survey topographical base map of 6 inches to a mile or 1:10 000 scale. I think that the use of such maps for drawing up 'final copy' geological maps is a mistake for the following reasons:

1. Thesis preparation by an undergraduate should be regarded as part of the training for future employment. Maps accompanying journal articles or company reports rarely use already published topographical maps as base maps. Such maps are normally completely redrawn and it is then possible to exclude all unnecessary topographical detail so that the geology shows up clearly. It is useful experience for students to draw their own maps from scratch, they will then appreciate some of the problems. The map should be easy to interpret, the geology should show up clearly, and there should be sufficient topography not only for structural interpretation but for another geologist to be able to use the map in the field and locate the positions of outcrops (Figs 40, 49, 52, 54, 63, etc.).
2. The published topographical map may not be at the most convenient scale for the geology. For example, a scale of 1:10 000 is far too small for some of the complex structural areas of Britain, and in some other countries the largest scale topograpical map available may be 1:25 000 or 1:50 000, again far too small if detailed mapping is contemplated.
3. Because of deficiencies in topographical maps the geological mapping may have been done entirely on aerial photographs at different scales from those of published maps (e.g. 1:7000 for many parts of Britain), and it is then more convenient to prepare the final map at this scale.

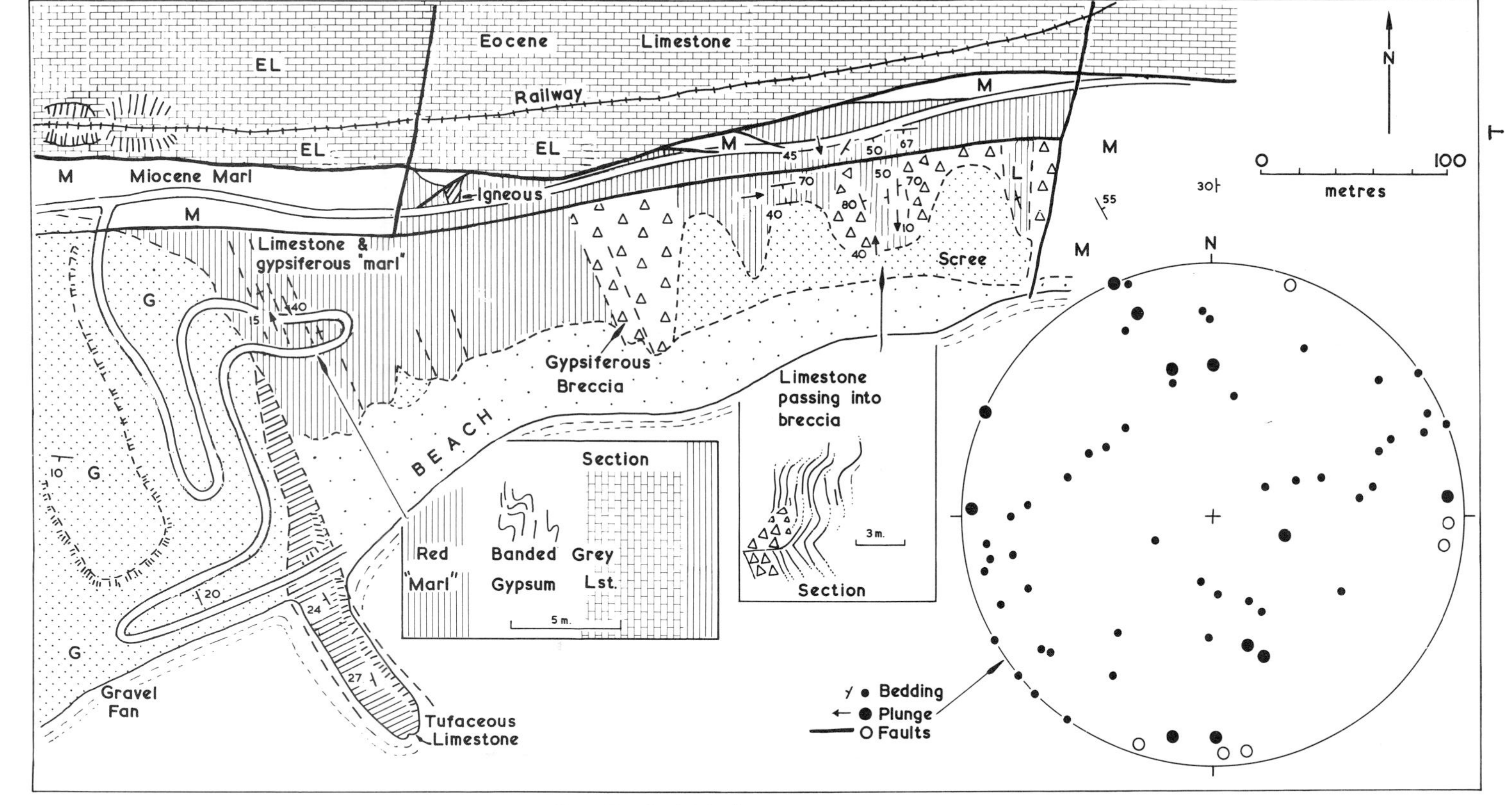

Fig. 38 A published enlargement map of part of a larger area, Sierra Bernia, SE Spain (Moseley 1973) to show important details for which the main map was too small in scale. Insets show even greater enlargements of some exposures; all the structures are plotted on a stereogram. Triassic limestone and gypsiferous marl and breccia (part of a diapiric intrusion) are faulted against Miocene and Eocene rocks. Quaternary tuffaceous limestone and gravel (G) is seen to have a tectonic dip, showing that the region is still tectonically active.

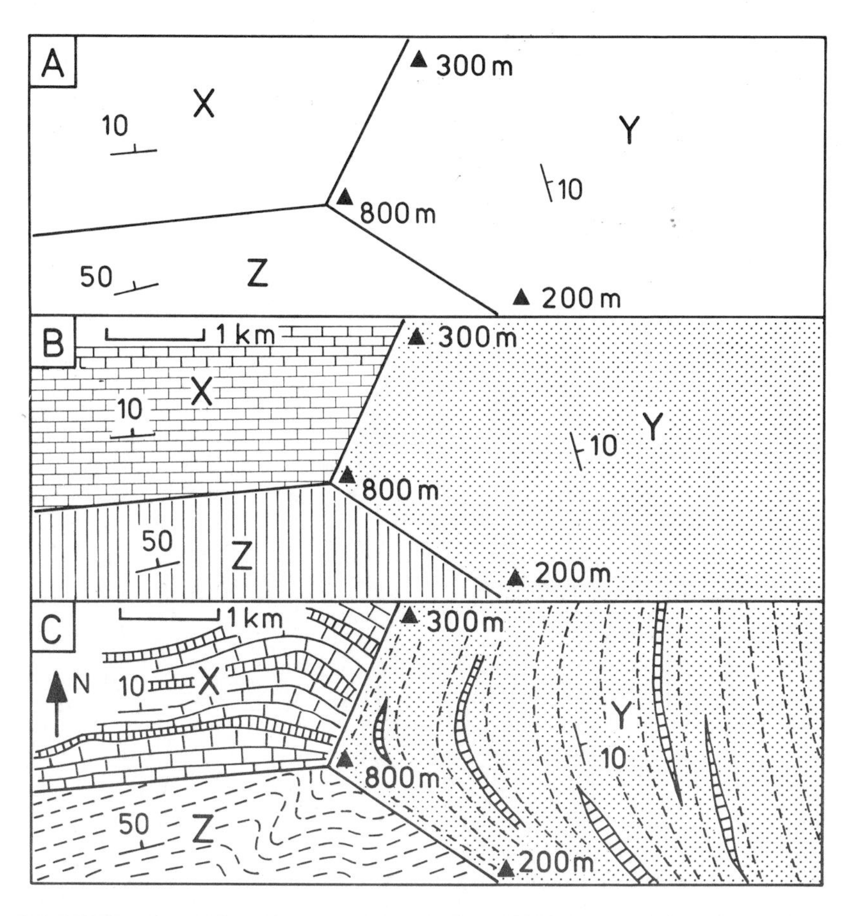

Fig. 39 Three maps of the same area to illustrate use of ornament. In **A** no ornament is used and the relations of rock units X, Y and Z are not clear. The standard adhesive ornament used in **B** does not help much in this respect, although it does separate the rock units clearly. In **C** the ornament is related to the geology to show outcrop and strike and the geological structure becomes apparent. Y, Sandstone with occasional interbedded shale; X, marine bedded limestone with shale bands; Z, predominantly shale. Note that maps should have a north point and a scale; both are shown on **C** but neither on **A**.

4. If the thesis is partly regarded as training for publication, it is a good idea for students to learn the requirements for maps that will be acceptable to geological journals (Figs 40, 52, 63, etc.).

In addition to folding maps and single-page maps some text figures are likely to be diagrams of exposures. It has been indicated in chapter 7 (page 40) that both photographs and field sketches are necessary in field notes, but that they each have their limitations, for example, compare Figs 34 and 42 to 44. The best type of illustration in most cases will be that which combines the best qualities of both. There are several methods of achieving this; perhaps the most common is to trace the important geological features from the photograph or to draw them from a projected transparency as described in chapter 8 (page 50) above, adding details from field sketches and incorporating structural and other information from field notes (Figs 44 and 95). Another simple method is enhancement of photocopies of the original photograph, on to which details of field sketches are incorporated (Figs 43, 56 and 112). The latter method has the advantage of preserving

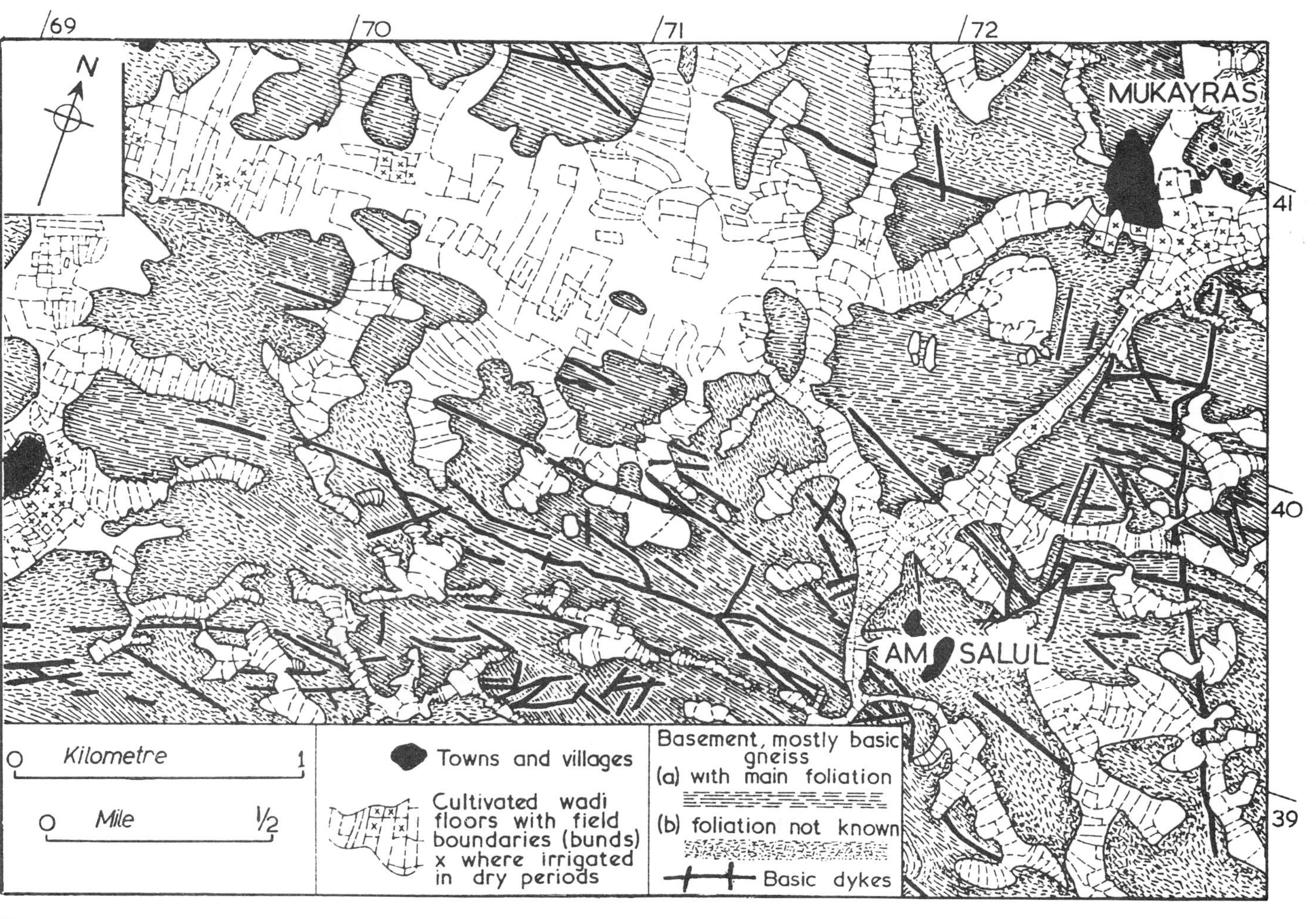

Fig. 40 Geological map of basement rocks in the Mukayras area of South Yemen (Moseley 1971a). The ornament on this map was entirely hand drawn, the advantage being that variations in foliation trend can be shown. These were mostly plotted from aerial photographs.

photographic detail, a near impossibility when photographs are traced by 'non-artistic' geologists (see page 50). It is worth repeating that the photocopy is modified using white ink (process white and a fine brush is the most effective) and black ball-point and felt-tip pens. The shadows and irrelevant details are painted white, the geological features are emphasized in black, and the result is photocopied again. The same principle can be applied with photocopied photogeological maps (see Figs 48, 49 and 116).

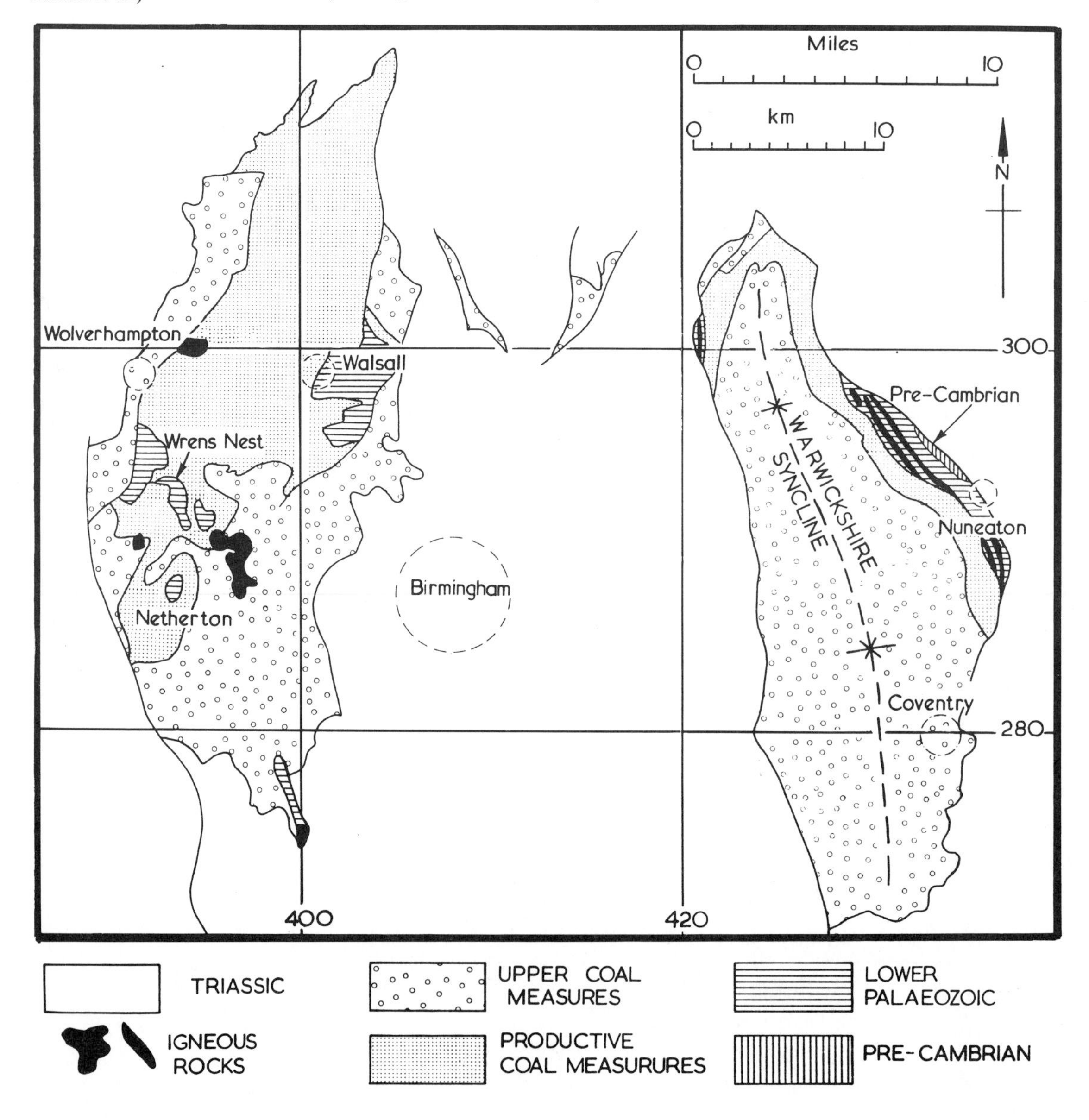

Fig. 41 Geological map of part of the English Midlands on which printed adhesive ornament has been used. On maps of this type it is best to leave one unit (usually the largest outcrop) as blank (white) and to make another (the smallest outcrop) black. The Triassic and igneous rocks are so represented on this map (see also Fig. 39). (After Moseley and Ahmed 1973.)

Data plotting on rose diagrams and stereographic projections

It is a fond illusion that if a sufficient number of field measurements are taken they will provide statistical data that can be plotted to give reliable estimates of orientations of cleavage, joints, bedding, fold axes and other structures. In the first place field measurement itself can be suspect. For example, if several individuals each measure 100 joint orientations in the same quarry significantly different results will be obtained. This is because different selection criteria are used. Some geologists will be influenced by the large master joints and others by the more numerous minor joints. It is not easy to think of field methods free from bias, but assuming that measurements are taken to the best of one's ability there remains the problem of how to plot the results (see Firman 1960).

Rose diagrams are particularly suitable for representing trends of a large variety of structures including sedimentological features such as current directions, and indeed many other aspects of geology can be shown in this way. Some of the hazards in their use can be demonstrated by referring to high-angle structures. For statistical validity it is important to have no less than 100 measurements, but this will depend on the size of the area, degree of exposure and the degree of concentration of structures. Cleavage measurements may concentrate within 10 or 20° whereas joints are usually much more diffuse. Some of the problems in plotting are illustrated by Figs 45 and 46. After analysis of several thousand of my own measurements I detected a distinct bias to record a trend of 360° rather than 359°, of 010° rather than 009°, and so on. Other observers may find they have similar habits which may go undetected. On rose diagrams it is usual to plot readings in 10° blocks and Fig. 45 shows what happens if one chooses 351 to 360; 001 to 010 rather than 350 to 359; 360 to 009 as the divisions. Another problem concerns the size of area to be represented on one diagram. Fig. 46 represents a plot of joint trends. It will be noticed that from south to north the structures swing in a gentle arc (D) and that two joint sets are represented (B and C). Yet if all are combined on to one diagram a false complexity results (A) and this must be guarded against. This same misleading representation is common on contoured stereographic projections and many such have been published. Figs. 49 and 66 clearly show how strong trends at many single localities become diffuse when all localities are plotted on one diagram. Fig. 49 also reveals that measurements plotted from aerial photographs do not necessarily conform exactly with ground measurements and should be compared with Fig. 48. Important conclusions must not be drawn from minor maxima on these general diagrams.

Other elementary mistakes commonly seen in theses (PhD as well as undergraduate) are: (1) Contoured stereograms are presented at 1%

Fig. 42 One of a stereopair of photographs of folds in Silurian greywackes and mudstones, Shap Fell, northern England. A sketch of the syncline for use in the field is shown on Fig. 34. Fig. 43 shows an enhancement drawing of a photocopy of this fold, and Fig. 44 is a published line drawing.

intervals when the number of points on the projection is less than 100; 1% contours erected for (say) 40 points is an obvious absurdity. (2) Contours which reach the edge of the stereogram are not continued on the (180°) opposite edge. (3) Lineations are plotted as poles. This shows no understanding of the three-dimensional geometry of the stereographic projection.

Reference to rocks, minerals, fossils and the literature

There are numerous pitfalls to avoid when describing and identifying rocks, minerals and fossils, and there are internationally agreed conventions of reference, especially for fossils and for citation of literature. It is important for undergraduates to be aware of these matters, and to pay attention to them when writing a thesis.

Rocks and minerals

There may seem to be few problems in describing and identifying rocks and minerals: difficult and unusual specimens can be brought back from the field and examined at leisure; thin sections and microfossil preparations can be studied. But all this can lull a student into the false security that the identifications are reliable. I will cite a few examples to show that this may not be the case.

1. Many volcanic rocks are identified both in theses and in many published articles specifically as rhyolite, dacite, andesite, etc. Most of these rocks have fine-grained altered matrices which certainly cannot be properly identified under the microscope, and they are frequently named in BSc theses according to the composition of the phenocrysts.

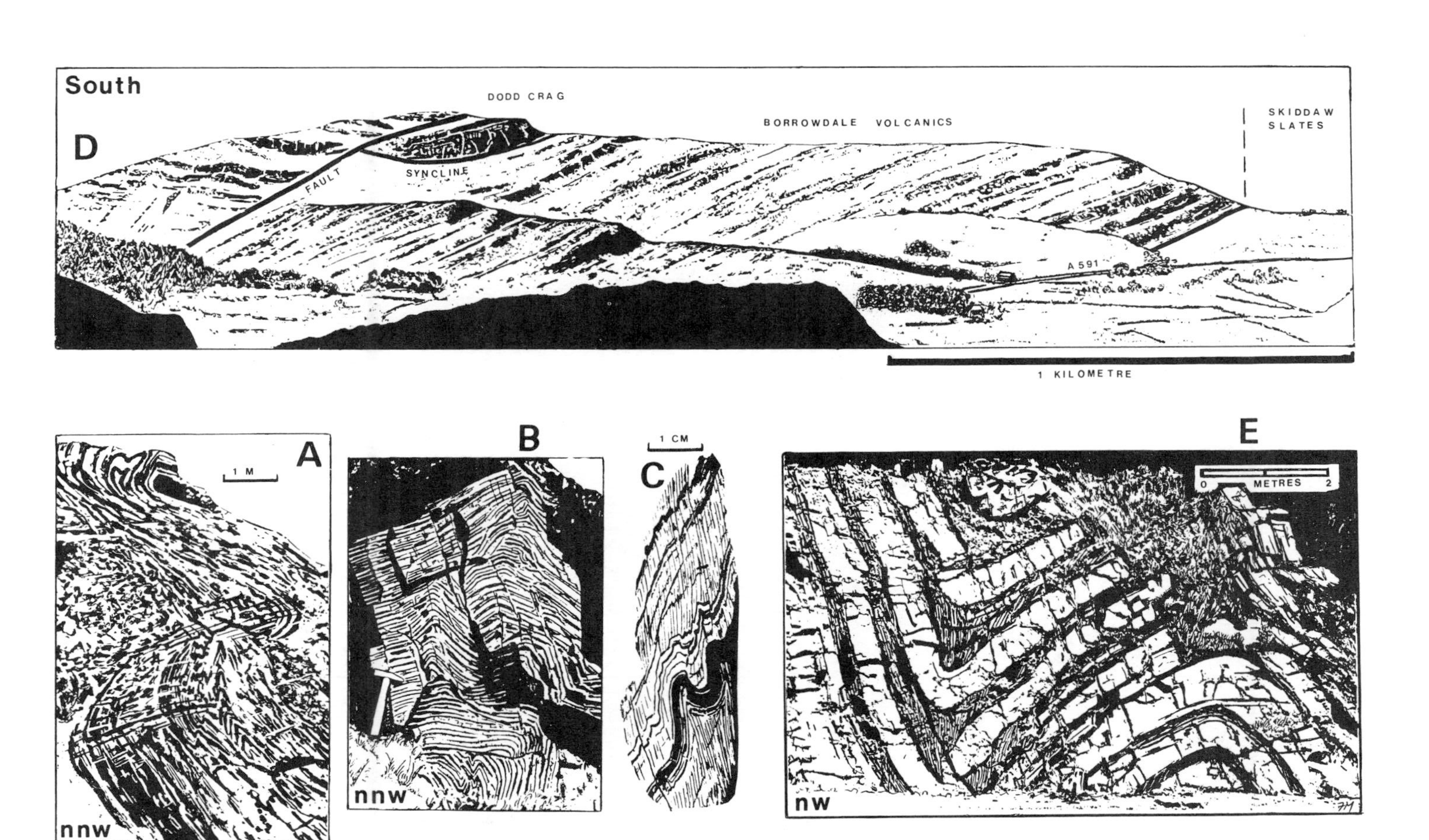

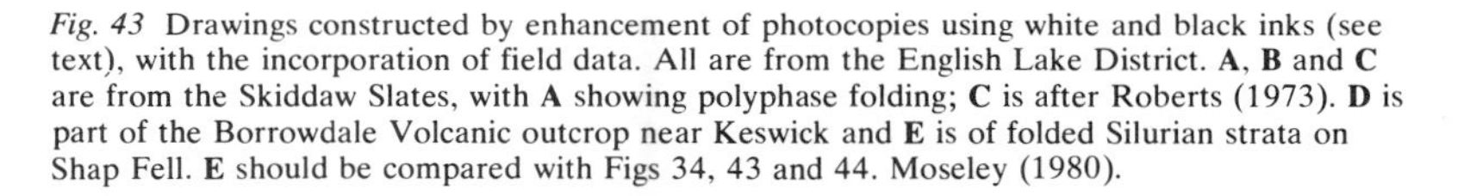

Fig. 43 Drawings constructed by enhancement of photocopies using white and black inks (see text), with the incorporation of field data. All are from the English Lake District. **A**, **B** and **C** are from the Skiddaw Slates, with **A** showing polyphase folding; **C** is after Roberts (1973). **D** is part of the Borrowdale Volcanic outcrop near Keswick and **E** is of folded Silurian strata on Shap Fell. **E** should be compared with Figs 34, 43 and 44. Moseley (1980).

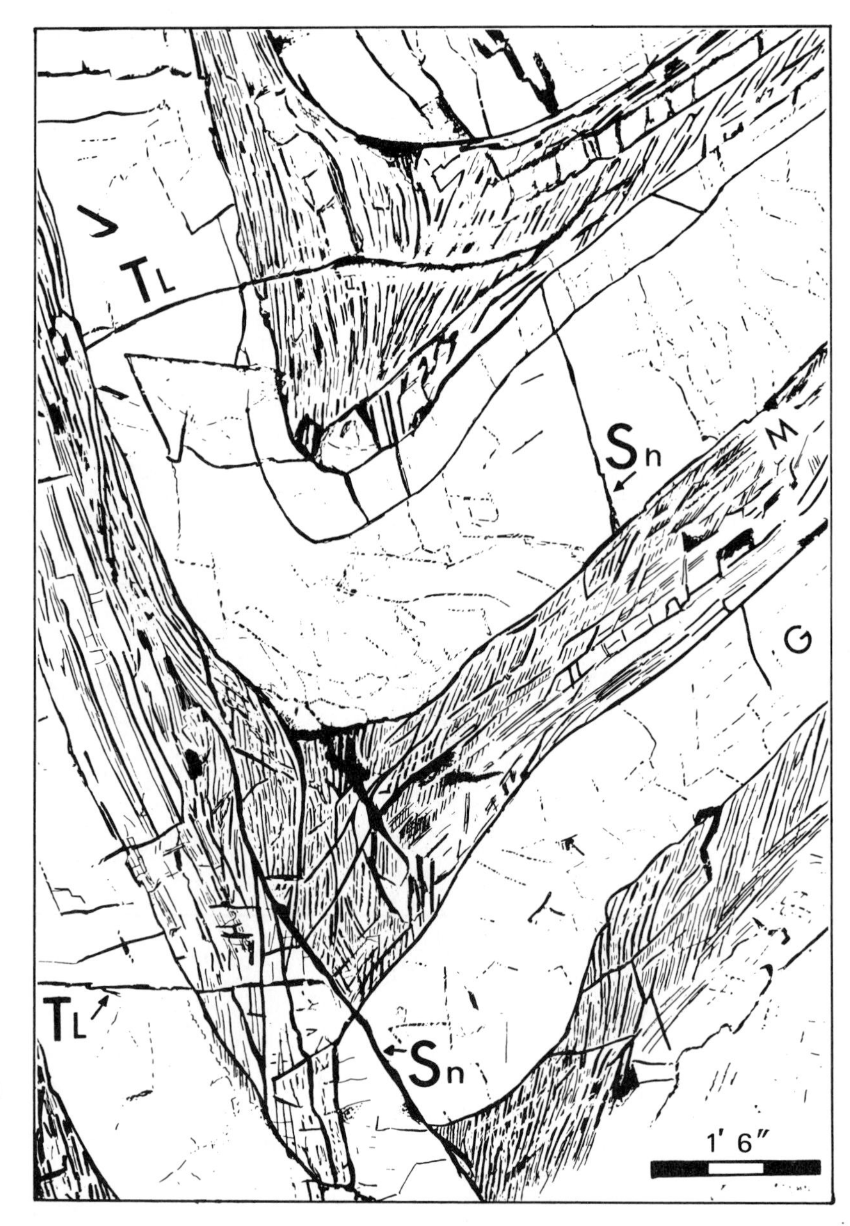

Fig. 44 Published line drawing of the syncline shown on Figs 34, 42 and 43. Use was made of the field drawing (Fig. 34) and of the stereophotographs. The effect is different from the alternative way of drawing structures shown on Fig. 43. (After Moseley 1968a.)

Such identifications are suspect and the precise name can rarely be justified without chemical analysis. These are not usually available to undergraduates.

2. Some minerals occur as fine-grained aggregates: for example, the alteration products chlorite and serpentine which may be loosely

named following casual inspection. One cannot be certain of their identity.

3. Augite is another mineral frequently mentioned by name in descriptions of thin sections and yet it is impossible to distinguish between this mineral and some other clino-pyroxenes by the optical methods normally used by undergraduates. There is therefore a danger in false precision and 'over identification' is not a sign of competence.

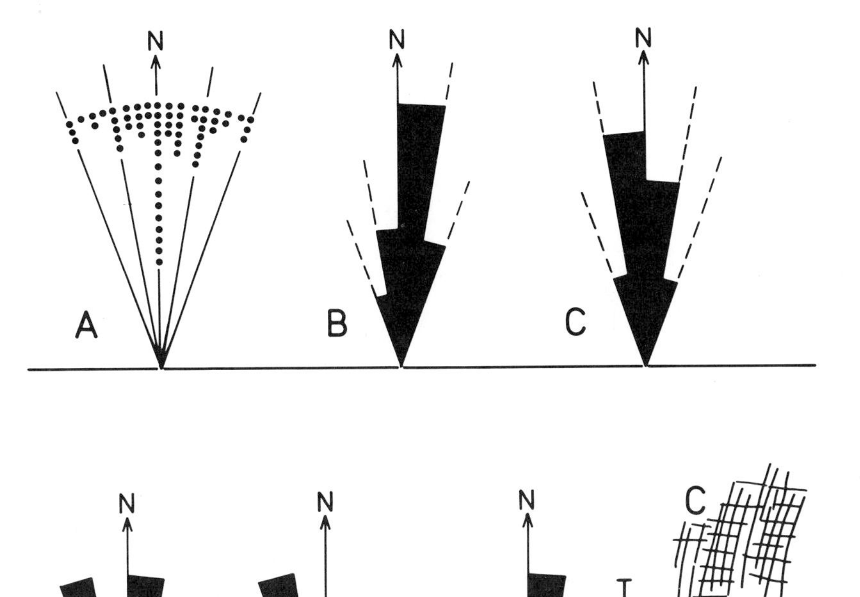

Fig. 45 Anomalies in data plotting on rose diagrams. **A**, Raw data of joint trends in which personal bias has resulted in more measurements to 360° (north) than to 359° etc.; **B**, rose diagram divisions of 350°–359°, 360°–009° etc.; **C**, rose diagram divisions of 351°–360°, 001°–010° etc. Note that **B** and **C** using the same data give different trends.

Fig. 46 Plots of joint trends for the area shown to the right. **A**, All measurements are plotted on one diagram; **B**, measurements for the south; **C**, measurements for the north. Diagram **A** gives a false impression of four joint sets, when in fact there are only two joint sets with a trend swing from south to north.

Fauna and flora

The fauna and flora should be referred to in the internationally accepted style. Before specific names are used it must be ascertained that all the diagnostic features relevant to that species are preserved. Many specimens are not sufficiently well preserved to permit specific identification; for example, many crushed shale specimens of graptolites and goniatites. Once it is decided that specific names are justified they should be referred to in the text in a standard way; for example,

Fig. 47 Carboniferous Limestone pavement (northern England) showing clints and grykes. The latter are eroded along well-defined joints, not always obvious at close quarters, but easily seen on aerial photographs (see Fig. 48).

'. . . the lower bands yield *Gothograptus nassa* (Holm) and *Pristograptus jaegeri* Holland *et al.*'. The brackets indicate some change in generic nomenclature subsequent to the first description of the species. The fossil names are printed in italics in journals, but should be underlined in typewritten accounts. There are some occasions when author names are not required; for example, ' . . . the mudstones contain a *Didymograptus hirundo* fauna', thus referring to the whole assemblage. Faunal assemblages from particular beds are usually listed as follows:

Anthracoceras of *paucilobum* group
Dimorphoceras sp.
Brachycycloceras koninckianum (d'Orbigny)
Brachycycloceras dilatatum (de Koninck)
Protocycloceras sp.
Tylonautilus nodiferus Armstrong

References to literature

References are also governed by internationally agreed rules, and abbreviations should conform with those in an accepted list of serial publications (such as the *World List of Scientific Periodicals*). If in doubt follow the style used in the major geological journals (although not all are identical in this respect). The reference list in this book can also be used as a guideline. It is generally the case that only works referred to in the text of a paper (or thesis) should be listed in the references. Articles referred to in the text should appear as follows: (1) according to Knarf (1976) the dinosaur beds are 2 cm thick; (2) the dinosaur beds are reported to be 2 cm thick (Knarf 1976); or if one wishes to be precise (Knarf 1976, p. 23). Where there are more than two authors the citation should be (Knarf *et al.* 1976), although all the authors should be referred to in the reference list at the end of the thesis.

Layout

The layout of a written account and the type of information in each section are most important. Concerning the latter it is necessary that 'factual' observation should be kept quite separate from interpretation and hypothesis. Reliable observers can record valuable data which can be used by others at a later date, even though their interpretation of it may be wrong. This is not possible if 'fact' and interpretation are intermingled.

The layout of the account is best decided by inspecting published papers on similar topics. It will obviously depend on the subject of the thesis. The example given below is of a stratigraphical topic, quite common in undergraduate theses.

The geology of the Upper Carboniferous rocks between Fagley and Shipley, Yorkshire

by J. Beckbottom

Abstract
(approximately one page)

1. Introduction
2. Stratigraphy
   (a) The Middle Grits
   (b) The Rough Rock
   (c) Lower Coal Measures
3. Palaeontology
4. Sedimentary petrology
5. Structure
   (a) Folds
   (b) Faults
6. Conclusions
7. References

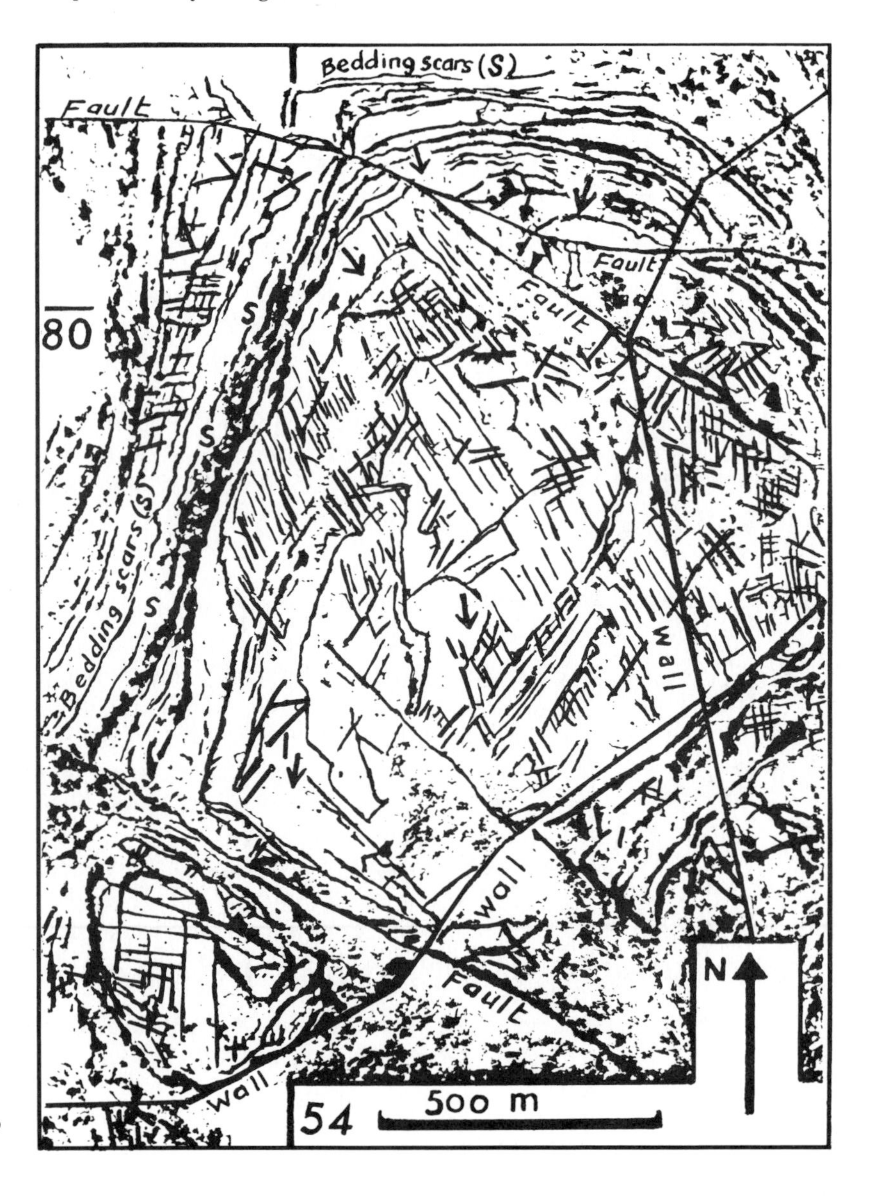

Fig. 48 Photogeological map of the Carboniferous Limestone outcrop of Hutton Roof near Lancaster, NW England. It was drawn by enhancement of a photocopy of one of the stereophotographs. Most of the area is exposed limestone pavement on which joints, bedding features and faults are easily seen. One method of measuring fracture orientations from aerial photographs is to place a 1-cm grid on the photogeological map (using a transparent overlay). The strikes of all fractures as they cross each square are then measured, long master joints thus being counted several times, and weighted against smaller joints. Care is necessary in mountainous terrain (see Fig. 4). Grid lines are marked on the margin of this map for comparison with Fig. 49.

Fig. 49 Published structural map of the Carboniferous Limestone of Hutton Roof near Lancaster, NW England. Compare with Fig. 48. Note that the joint measurements from aerial photographs correspond approximately but not exactly with those made on the ground (rose diagrams **A** and **B**) (Moseley 1972a). (© Geological Society of London.) ▷

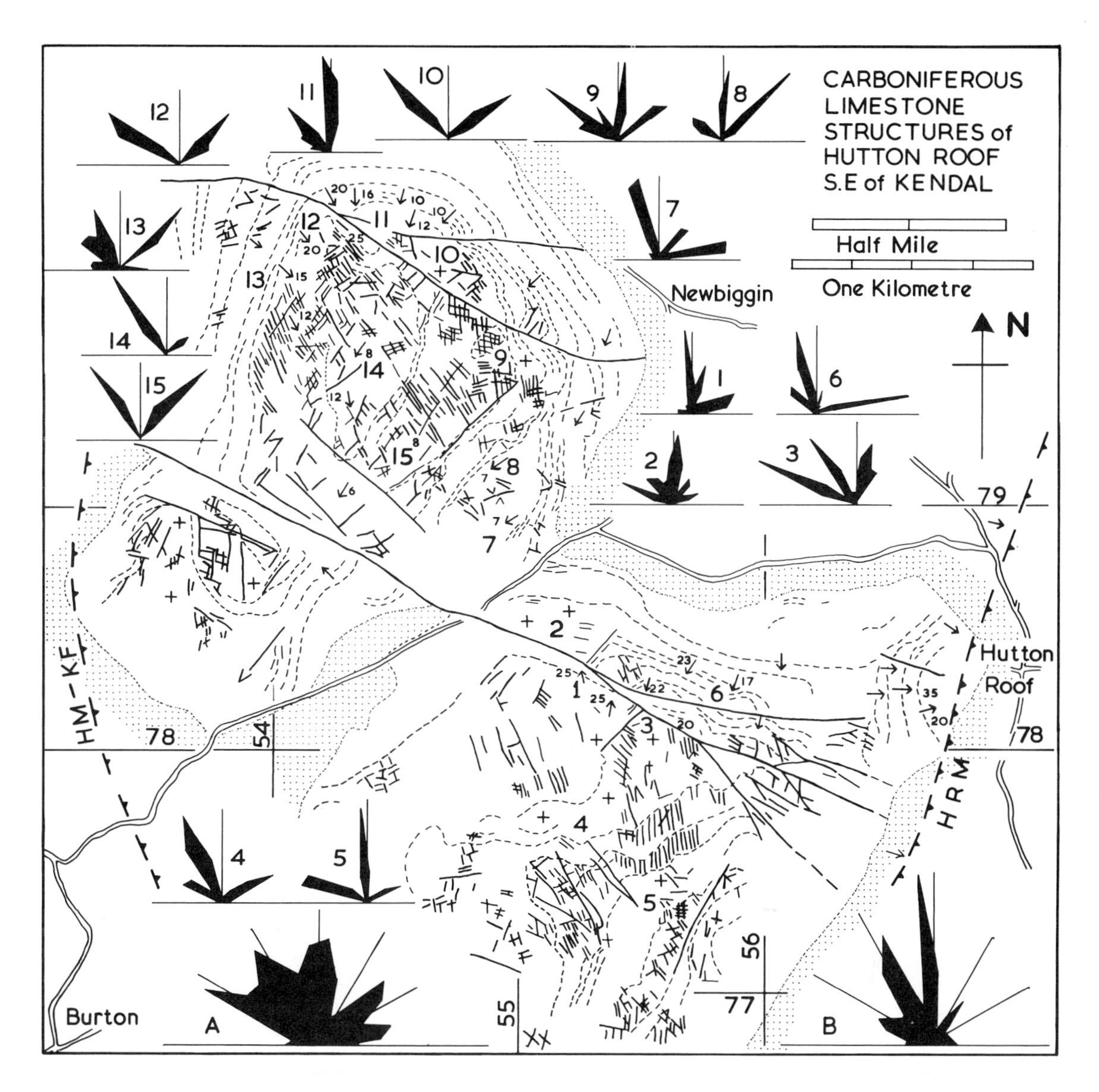
CARBONIFEROUS LIMESTONE STRUCTURES of HUTTON ROOF S.E of KENDAL
Half Mile
One Kilometre
N
Newbiggin
Hutton Roof
Burton
HM-KF
HRM
78
79
77
54
55
56
A
B

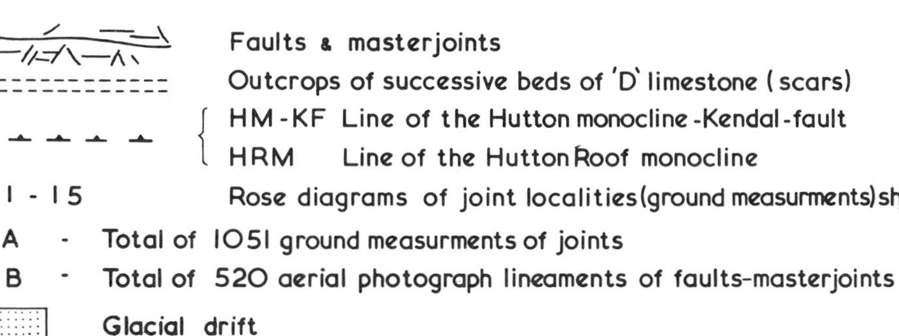
Faults & masterjoints
Outcrops of successive beds of 'D' limestone (scars)
HM-KF Line of the Hutton monocline-Kendal-fault
HRM Line of the Hutton Roof monocline
1 - 15 Rose diagrams of joint localities (ground measurments) shown on map
A - Total of 1051 ground measurments of joints
B - Total of 520 aerial photograph lineaments of faults-masterjoints
Glacial drift

Subdivision of this kind makes the writing much easier, since it can be written section by section. It also makes reading and cross-reference easier, an important factor when a thesis is being examined for a university degree. Section 1 would normally contain such items as the history of research, with reference to adjacent areas and comparable problems elsewhere; sections 2–5 would be factual records of the geology; and section 6 would be a discussion of results with interpretations and hypotheses.

Since most undergraduate theses are reports of geological mapping projects, there will be a geological map, or maps which should be folded and inserted in a pocket at the back of the thesis. Long cross-sections can be dealt with in the same way. Sometimes maps are submitted separately in a roll, but this is frustrating to the reader as the map tries to reroll itself. Final presentation is probably most effective if the thesis is provided with covers of thin card upon which there is a title, and is then made into a neat book using spiral binding or some similar method. It is a mistake to submit loose-leaf pages in a file or in a spring-back file; beautifully documented accounts presented in this way can easily be accidentally dropped during examination when they may spread themselves over a large area of office floor!

The ultimate layout of an undergraduate report or postgraduate thesis is the responsibility of the student concerned. This will not always be the case in postgraduate employment, however, since most government surveys and private companies have definite rules on layout and presentation which have to be strictly adhered to.

PART II Case Histories

11 Carboniferous sedimentary rocks in the Pennine Uplands of northern England

The greater part of the Pennine hills is composed of faulted and gently folded or tilted sedimentary rocks of Carboniferous age, which have worldwide analogues where similar techniques of survey could be practised: for example there are similar sequences and structures in the Pennsylvanian of the north-east of the United States; in New South Wales and Queensland; and in other Carboniferous regions of NW Europe. The sequence in the Pennines is of variable thickness, approaching 6000 m in the Pennine Basin, and is subdivided into a lower division of the Carboniferous Limestone, followed by the Millstone Grit and the Coal Measures. A further division is the Yoredale Group, a facies change in the north and the same age as the upper part of the Carboniferous Limestone and lower part of the Millstone Grit of regions further south.

Apart from the Coal Measures, which often occupy lower ground, most of these rocks outcrop on open rolling uplands of moorland and poor pasture, with a relief of up to 500 m. There is deep incision by streams of the present erosion cycle, and valley-in-valley forms are common. Access is generally good and it is always possible to travel by car to within 1 or 2 miles of any outcrop.

This account deals with geological survey of parts of the Millstone Grit and Yoredale Groups, for which field methods are similar. Of the other divisions, the Carboniferous Limestone has more uniform lithology that requires the attention of experts in carbonate petrology; the Coal Measures tend to be badly exposed, although there is an abundance of mining information. Further information on the areas described will be found in Burgess and Wadge (1974), Moseley (1954, 1956 and 1962) and Shotton (1935).

The Millstone Grit and Yoredale Groups each consist of alternations of hard and soft rock; predominantly sandstones and shales in the former, and sandstones, shales and limestones in the latter. The hard rocks resist weathering to form escarpments and mapping features, whereas the shales weather back to form parts of the separating benches and lower parts of the escarpments, and in these situations are poorly exposed. Other rocks forming minor parts of the sequences include thin coal seam and their attendant seat earths, generally ganisteroid sandstone. Exposures of all the rock types are good, however,

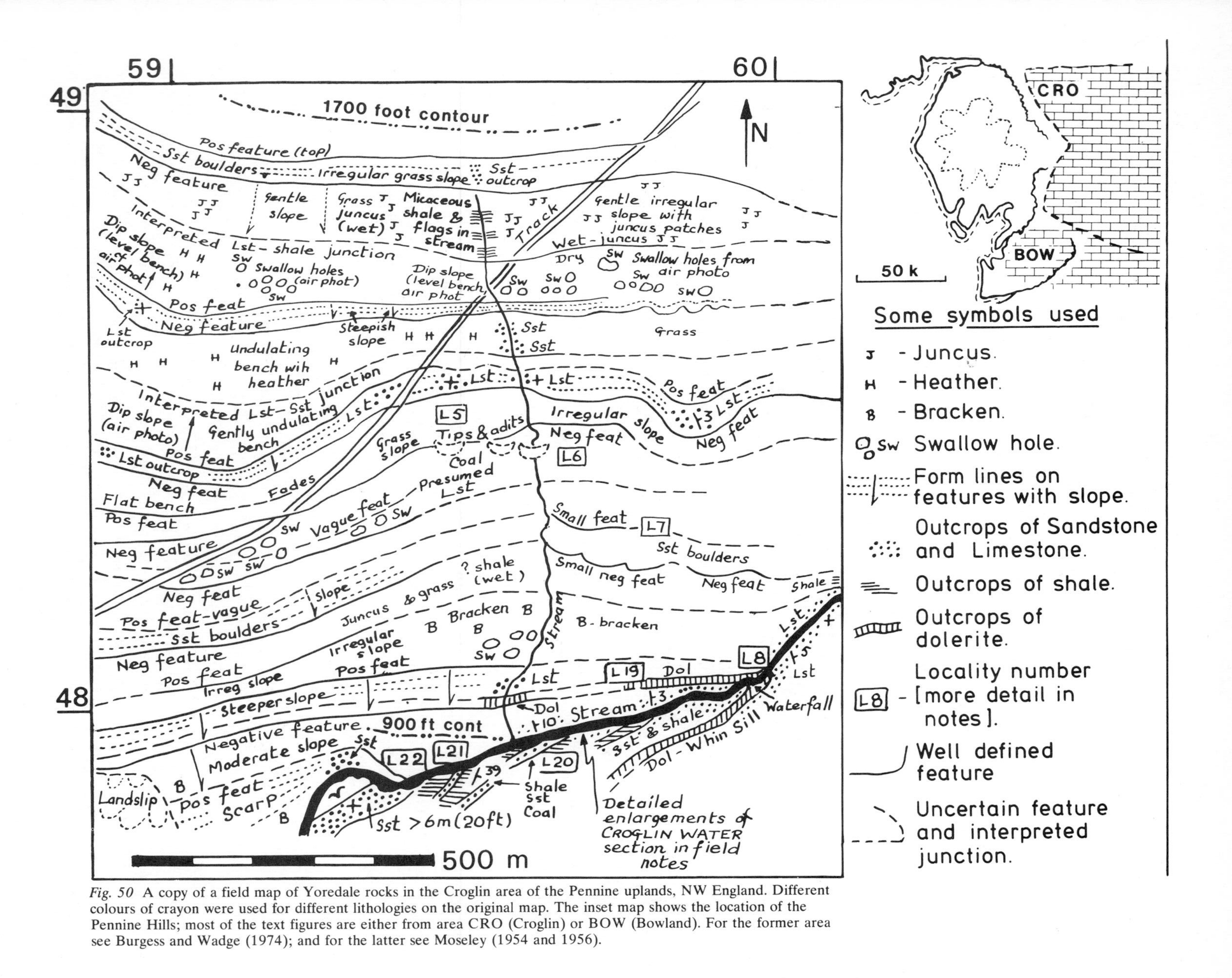

Fig. 50 A copy of a field map of Yoredale rocks in the Croglin area of the Pennine uplands, NW England. Different colours of crayon were used for different lithologies on the original map. The inset map shows the location of the Pennine Hills; most of the text figures are either from area CRO (Croglin) or BOW (Bowland). For the former area see Burgess and Wadge (1974); and for the latter see Moseley (1954 and 1956).

along the incised valley sides and stream bottoms, and frequently these valleys form almost continuously exposed sections. These two facets form the basis of geological survey; the details of sequences are established along the valley bottoms, and the harder bands (sandstone and limestone) are then mapped from one valley to the next by making use of escarpments and features. Survey procedures are as follows.

Use of topographical base maps

Most of Britain is covered by high quality topographical maps of 6 inches to 1 mile or 1:10 000, the metric scale which is replacing the 6-inch maps. There are also 1:2500 maps available for all the populated regions. Until aerial photographs became available these maps were used successfully for geological mapping, and for most of the region the most effective mapping method is still to plot geological information directly on to the maps rather than on to aerial photographs. These maps show all roads, tracks, footpaths, streams, buildings, field boundaries, etc. so that it is usually possible to locate position to within a few metres, even without use of a tape measure.

Use of aerial photographs

Even though the published maps are of such good quality, they do not show all those items useful to geological survey that are often visible on photographs, and photographs should therefore be used for the additional information they supply. They are particularly useful on the higher moorland areas where identifiable map features (walls, streams, etc.) may be 1 km or more apart and it takes much time in simply locating position if geological features are plotted directly on to a map. The air photographs allow these features to be located immediately, thus saving much time. The photographs also reveal vegetation changes that can be related to the underlying rock. Most obvious is the bracken and heather on drier ground (sandstone) and *Juncus* in marshy areas (usually impermeable shale). Swallowholes indicating limestone show up clearly on the photographs; it can take a long time to plot accurate positions of all swallowholes by ground survey alone, but their positions should be recorded. Consider, for example, their importance in the construction of a highway. Fault lines are also seen more clearly on the photographs than on the ground (Fig. 48), and even if the area is covered by glacial deposits fault structures often show through (Norman 1968 and 1970). However, aerial photographs can on occasions be misleading. Some vegetational patterns can change year by year. Irregular patches of heather may be burnt during a summer and the resulting dark patches seen on photographs are not representative of heather distribution at the time of survey. Also most of the detailed rock sequences will be located along stream bottoms, perhaps obscured by woodland; these will not be visible on the photographs, so one might get a false impression of badly exposed ground.

Survey methods

It has been indicated that survey proceeds mainly by mapping features and escarpments of the harder rocks across open ground, and by measuring the details of the sequences along the incised stream courses.

Escarpments and features

The harder rock—sandstone and limestone—is generally exposed in the escarpments, but many of the smaller features are completely covered by soil or a mantle of hillwash, so that the nature of the rock has to be inferred from the topography. Mistakes can be made in regions where there are glacial deposits since similar features can be formed by lateral moraine; and the soft shale is rarely exposed so it is easy to underestimate its importance. The difficulties of estimating thicknesses of rock formations from features is illustrated by Fig. 51. At the one extreme it is possible for the hard rock to be a thin layer capping the feature and protecting the softer rock below from erosion, whereas at the other extreme the hard rock may form the whole of the feature. Generally the problem has to be decided empirically by following the feature into a stream section and checking how it relates to the rock exposures there. Hilly regions that have been glacierized pose still more problems. Solifluxion and hillwash mantle nearly all the slopes, and it is the harder, more resistant rocks that dominate these subsoil formations. Since shale is friable and limestone subject to chemical weathering, this means that the more massive sandstones generally form the majority of fragments in this surface layer. A limestone formation may often be mantled by soliflucted sandstone boulders brought down from a higher level and it is an easy mistake to take such a feature for a sandstone outcrop. The clues to watch for are: (1) a limestone outcrop is likely to be dotted with swallowholes, recognizable even though they may be choked with sandstone boulders (Figs 50 and 51); and (2) limestone outcrops in these Pennine areas are often associated with short, but lush, green grass.

Should it be necessary to construct a section of strata from a region where escarpments and features are common but where there are few stream sections, it can be attempted as indicated on Figs 52 and 53. Such a section at best can only be an approximation. Construction will primarily depend on sketching in structure contours using the methods all students are introduced to on elementary geological maps. These

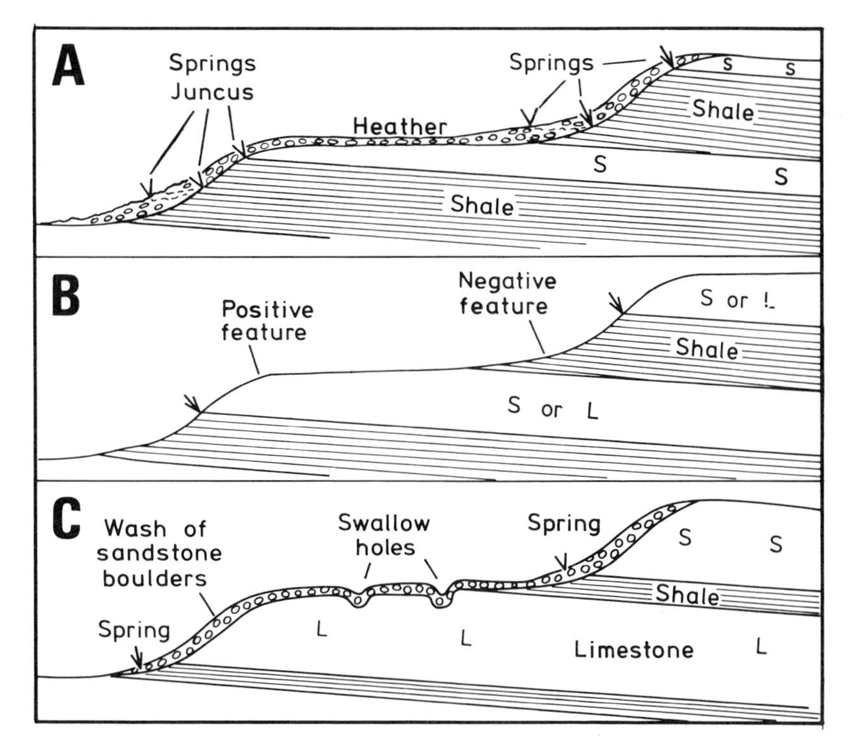

Fig. 51 Diagrammatic sections to illustrate the relation between lithology and topography in the Millstone Grit and Yoredale Groups of northern England.

A. The harder rock (S, sandstone) caps the escarpments with softer shale forming the slopes of the escarpments. The solid rocks are concealed by downwash, mostly of more resistant sandstone fragments from higher up the hillside. Springs initially located on sandstone shale boundaries may also emerge at other positions after seepage through the downwash. In this situation shale thickness may be underestimated.

B. The junction between harder and softer rock here occurs at the point of inflection on the escarpment. Note the positions of positive and negative features, likely to be plotted on a field map, but not necessarily the true junction.

C. In this case the hard rocks form the whole of the escarpments and shale is subsidiary, but the topography does not differ greatly from A. Hillwash conceals outcrops but limestone is indicated by the presence of swallowholes.

will be checked against measured dips, but it must be remembered that isolated dip readings can be unreliable in terms of a regional dip. There are frequent local flexures, valley bulge structures (see below) and undulations in bedding resulting from sedimentation and these often result in anomalous readings. There are also the difficulties mentioned above (Fig. 51), of knowing how much of the mostly unexposed section is to be interpreted as the hard rock and how much is the soft rock. Nevertheless the total thickness of strata, perhaps several hundred metres, can be more reliably estimated this way than by measuring up numerous disconnected sections along a stream course (below).

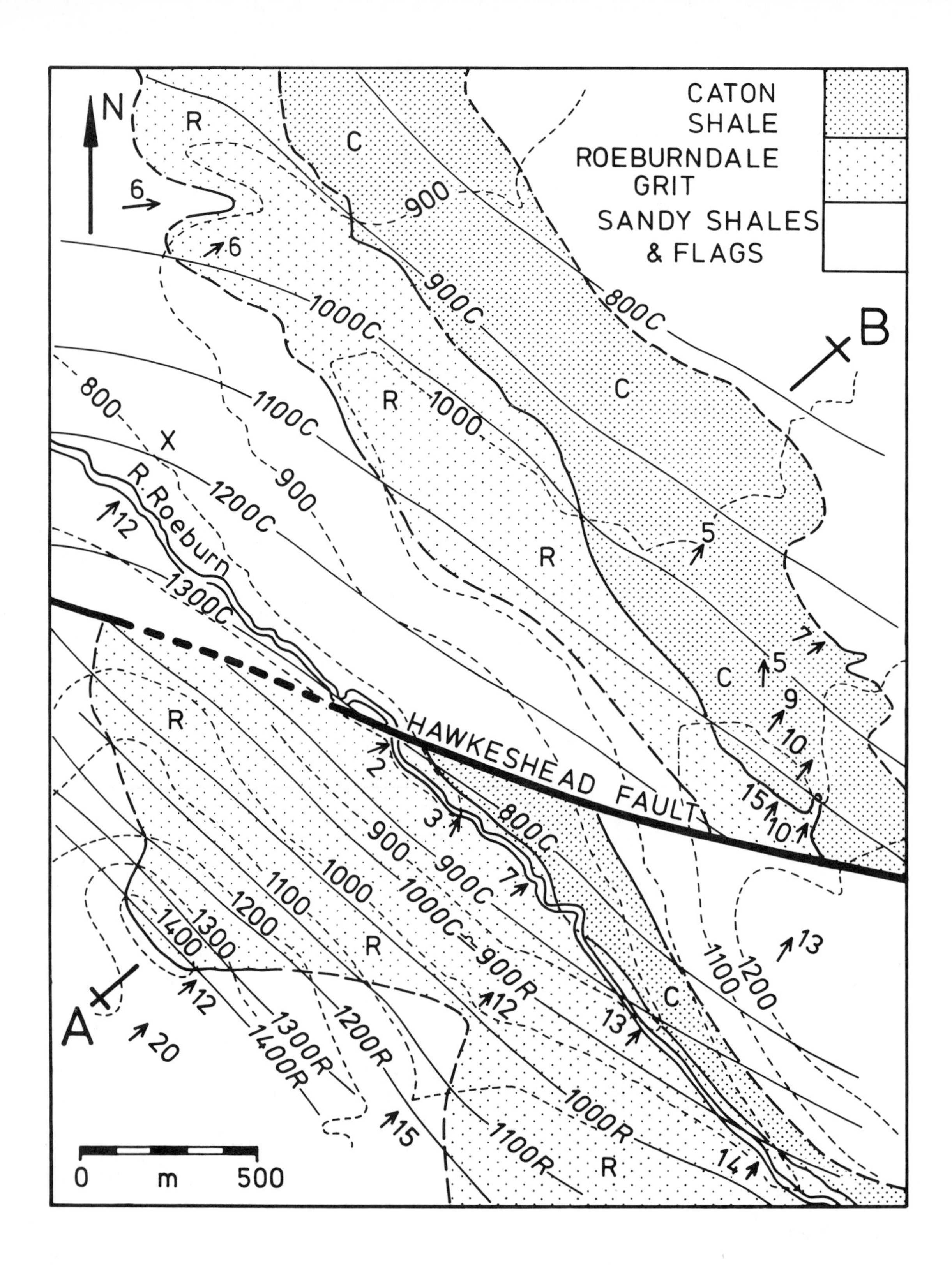

N
CATON SHALE
ROEBURNDALE GRIT
SANDY SHALES & FLAGS
B
A
R. Roeburn
HAWKESHEAD FAULT
0 m 500

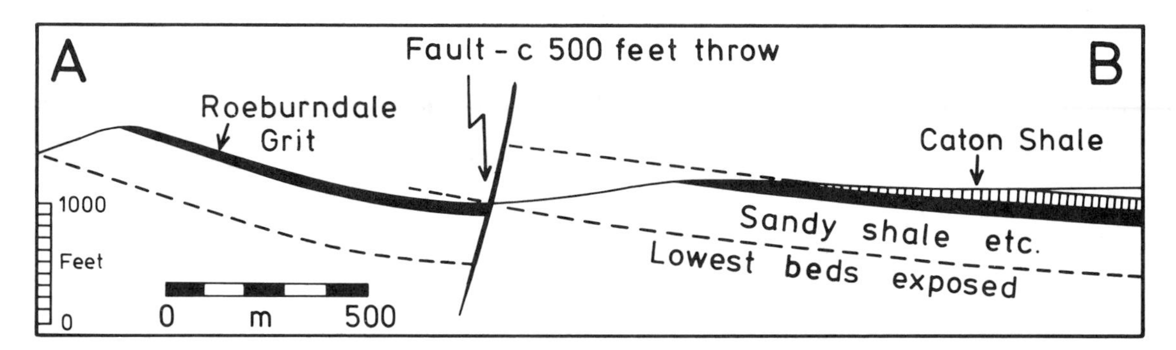

Fig. 53 Section across Fig. 52 along the line indicated. The structure contours on Fig. 52 were used in construction of the section.

Stream sections and quarries

Complementary to the mapping of escarpments and features are the records of exposures along stream courses and in quarries. The outcrops vary from almost continuously exposed sections, where the streams are incised into solid rock, to zero exposure, especially where there are thick deposits of drift. The streams and quarries indeed provide most of the information on which detailed understanding of the geology is based. Sequences can be examined foot by foot, sedimentology can be studied, the nature of junctions between formations ascertained, fossil horizons located, and faunal and floral assemblages collected, so that the sequences can be dated and correlated with adjacent sections and with other areas (Figs 25, 26 and 54).

Sedimentology

Any field investigation of these Carboniferous sedimentary rocks must concern itself with the nature and origin of the sediments. This is a rapidly expanding subject of study and requires careful, discerning observation of the numerous sedimentary features present in almost every outcrop. It is clearly most important that the surveyor should have consulted the literature (for example, the *Journal of Sedimentary Petrology* and, for northern England, the *Proceedings of the Yorkshire Geological Society* and the *Geological Journal*) before going into the field to discover the variety of sediment and structure likely to be encountered. Even so not all the observed features will be understood at the time they are recorded, and it is necessary to be as objective as possible and to make numerous field sketches accompanied by photographs.

◁ *Fig. 52* Map of part of the Millstone Grit outcrop of the Bowland Fells, NW England (Moseley 1954). In gently dipping sedimentary rocks of this type construction of structure contours facilitates estimation of fault displacement, determination of stratal thicknesses, and the construction of sections (see Fig. 53). Too much reliance should not be placed on individual dip measurements since minor undulations of bedding are common.

The Millstone Grit and Yoredale Groups (Supergroups) are predominantly clastic, with upward-coarsening units perhaps the most common, representing the outward growth of fluviatile and deltaic silts and sands into marine environments or into on-delta lakes. Each of these episodes is likely to end with on-delta accumulation of peat (now coal), thus completing a single cyclothem (typical of the Millstone Grit). The

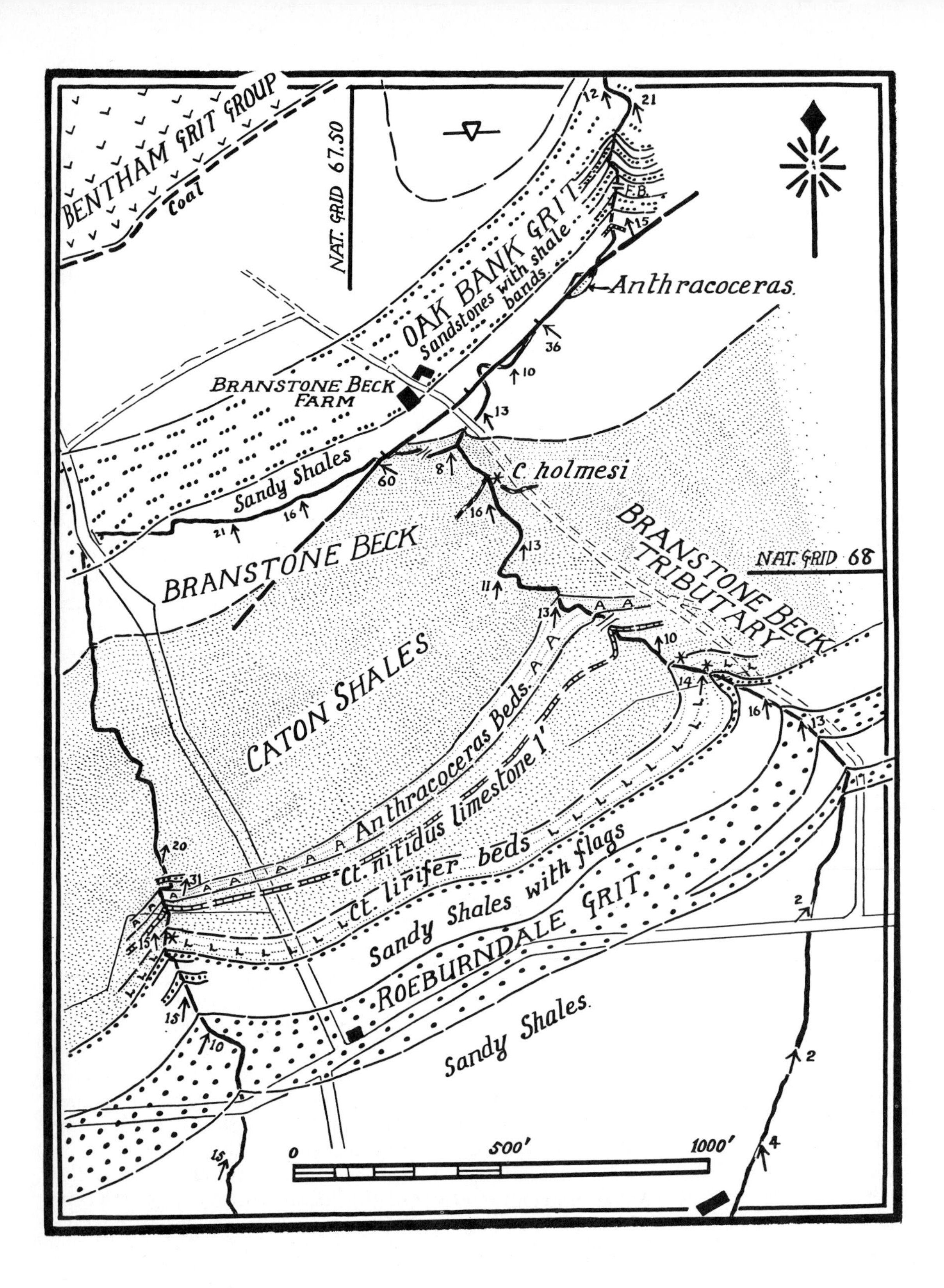

BENTHAM GRIT GROUP
Coal
NAT. GRID 67.50
OAK BANK GRIT
Sandstones with shale bands.
F.B.
Anthracoceras.
BRANSTONE BECK FARM
Sandy Shales
C. holmesi
BRANSTONE BECK
BRANSTONE BECK TRIBUTARY
NAT. GRID 68
CATON SHALES
Anthracoceras Beds.
Ct. nitidus limestone 1'
Ct. lirifer beds
Sandy Shales with flags
ROEBURNDALE GRIT
Sandy Shales.
Sandy Shales
0
500'
1000'

process can be repeated, with variations, many times, but the Yoredale cyclothem is different in one major respect in that it contains a limestone unit (Figs 50, 55 and 56). Detailed observations will derive from stream sections and quarries and will be extremely varied. They may include erosive-based sandstones, channel-fill deposits, with festoon bedding, scour hollows and intraformational conglomerates of various kinds. Cross-stratification is likely to be present on all scales down to the micro-cross stratification to be seen in siltstones, and current-ripple lamination is also found. Planar and wavy lamination is common, especially in siltstone, as is deformed stratification, which may include slump structures, convolute bedding, accommodation faults and load balls (Fig. 57). Delta-front channel turbidites are also found in some areas with good examples of erosive sole marks, especially groove casts. An understanding of all these structures will require consultation not only of the numerous papers on the Carboniferous of the Pennines, but also of descriptions of similar environments of the present day, such as the Mississippi Delta. There will also be many occasions when it will be possible to determine palaeocurrent direction; the best results are generally obtained from the larger scale cross-stratification (deltaic bedding). As many measurements as possible should be taken on the orientations of foreset beds, which can then be plotted statistically. Other structures such as groove casts should also be measured if they are present. If the regional dip (topset beds etc.) is low, the measurements can be plotted directly on to rose diagrams as indicated in chapter 10 (page 65) (Figs 45 and 46), but if the rocks have been tectonically folded they will have to be plotted stereographically and corrections made as indicated on Fig. 58.

Palaeontology

A large proportion of the Millstone Grit and Yoredale Groups are of fluviatile and deltaic origin. These sequences, being sparsely fossiliferous, generally yield no more than scattered plant fragments but there are occasional beds crowded with freshwater bivalves, especially near the top of the Millstone Grit. More importantly, there were also numerous marine incursions and the resulting marine faunas are of great value in zoning and in correlating sequences with those elsewhere. It will be obvious that the faunas should be recorded, identified and described. Marine shales in the Millstone Grit sequence have faunas dominated by goniatites and bivalves, whereas in the Yoredale limestones corals and brachiopods are the common fossils.

Although the Millstone Grit is predominantly continental in origin, there are a few marine shale formations exceeding 50 m in thickness, and numerous marine 'bands' which may be no more than 1 m thick. It does not require much field experience to distinguish between these marine shales, which are fine grained and almost black, and the grey,

◁ *Fig. 54* Geological map of a Millstone Grit area near Ingleton, NW England, produced as part of a PhD thesis (Moseley 1954). In areas such as this exposures are good along stream courses, where details of the successions are determined. Adjacent successions are then correlated by faunal and lithological comparison, and by mapping features from one stream to the next.

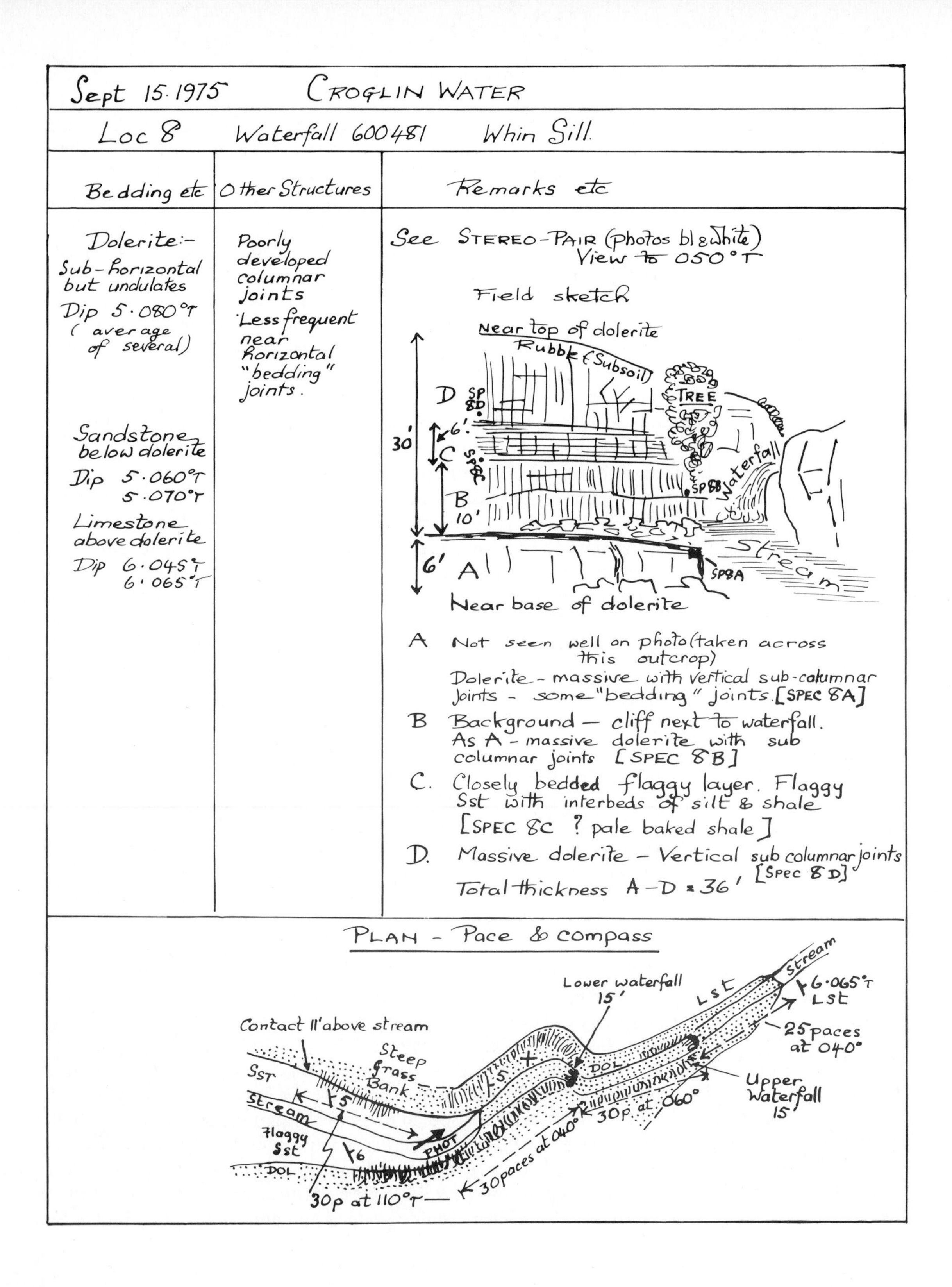

Sept 15 1975 CROGLIN WATER

Loc 8 Waterfall 600481 Whin Sill.

Bedding etc	Other Structures	Remarks etc
Dolerite:- Sub-horizontal but undulates Dip 5·080°T (average of several) Sandstone below dolerite Dip 5·060°T 5·070°T Limestone above dolerite Dip 6·045°T 6·065°T	Poorly developed columnar joints Less frequent near horizontal "bedding" joints.	See STEREO-PAIR (photos bl & white) View to 050°T Field sketch A Not seen well on photo (taken across this outcrop) Dolerite - massive with vertical sub-columnar joints - some "bedding" joints. [SPEC 8A] B Background — cliff next to waterfall. As A - massive dolerite with sub columnar joints [SPEC 8B] C. Closely bedded flaggy layer. Flaggy Sst with interbeds of silt & shale [SPEC 8C ? pale baked shale] D. Massive dolerite - Vertical sub columnar joints [Spec 8D] Total thickness A–D = 36'

PLAN - Pace & compass

micaceous, silty to sandy shales and mudstones of deltaic or fluviatile origin. The marine formations and horizons are most useful markers and should be carefully examined. Goniatites are especially important since they evolved rapidly and each horizon yields a distinctive faunal assemblage. They are usually found as crushed specimens, but ornament is well preserved and specific identification is often possible. As an example the Caton Shale sequence, in the middle of the Millstone Grits of the Lancaster Fells, is illustrated (Figs 36, 54 and 59). These shales, which are up to 60 m thick, have a uniform lithofacies but there are significant changes in fauna. First, there are bands, usually less than 1 m thick, which are rich in *Cravenoceras* and allied genera; these are separated by much thicker sequences dominated by *Anthracoceras*. These changes have been interpreted as evidence of alternations between truly marine conditions and estuarine, perhaps brackish-water conditions. Secondly the cravenoceratids show rapid evolutionary changes and several subzones can be traced over large parts of western Europe. The importance of spending time collecting representative faunas from both the thicker marine shales and the marine bands will therefore be apparent.

The Yoredale rocks, with their coral–brachiopod faunas, are faunally quite different from the Millstone Grit. The virtual absence of goniatites makes zoning and correlation of the rocks more difficult, but it is none the less important to record carefully the faunas, which are almost entirely within the limestones.

◁ *Fig. 55* Copy of a page of field notes for the Croglin area (see Fig. 50). The sketch of the Whin Sill was later drawn to scale using the stereophotographs (see Fig. 56).

Fig. 56 Drawing (enhanced photocopy of one of the stereopair indicated on Fig. 55) of the outcrop of the Whin Sill (dolerite) in Croglin Water, north Pennines. The Sill is split by a flaggy sandstone member.

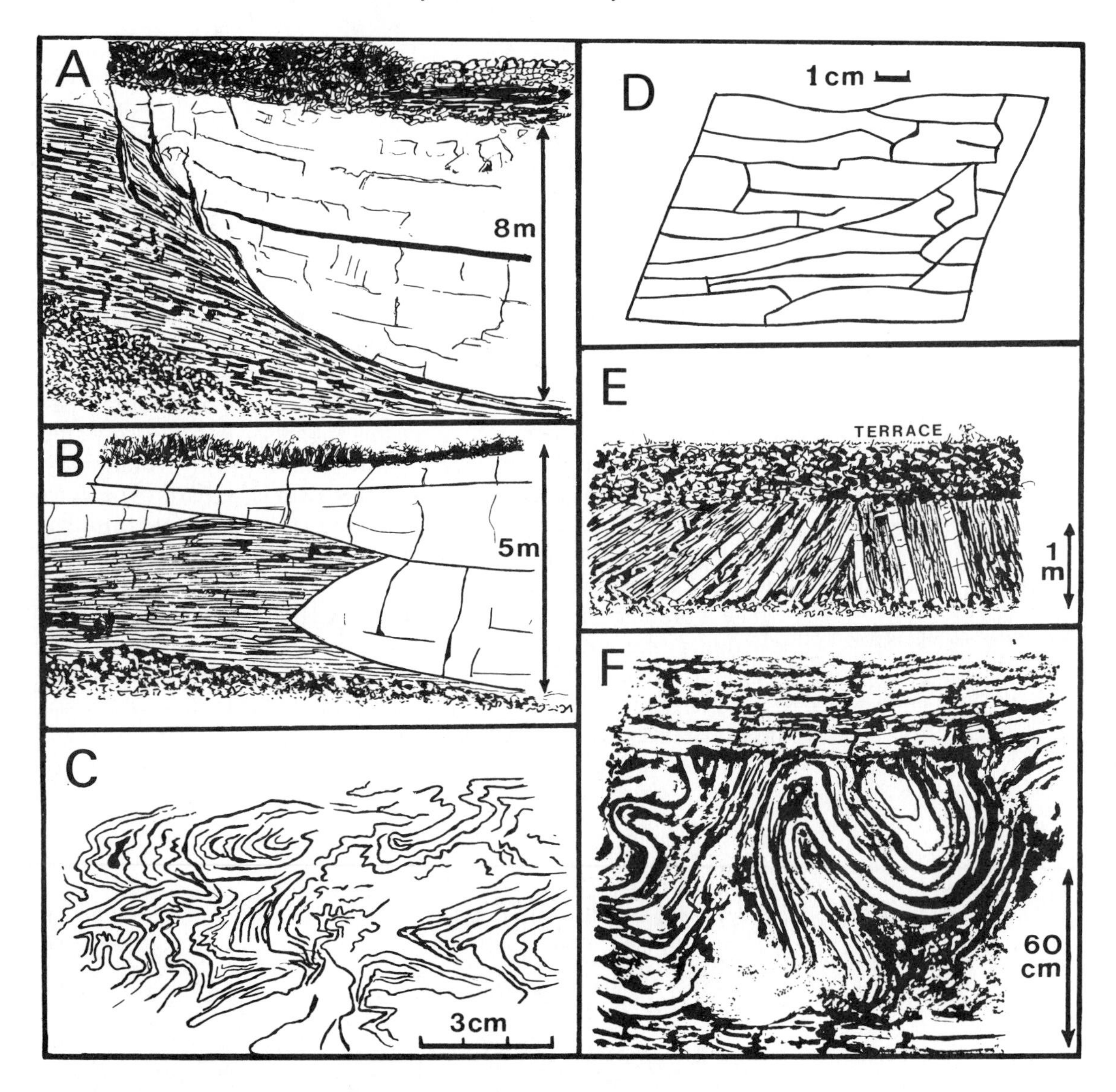

Structure

The Carboniferous rocks of the Pennines are generally not highly deformed (Figs 50, 52, 54 and 63) and consequently structural studies tend to form minor parts of most surveys. Nevertheless the importance of structure must not be underestimated. There are many occasions when failure to recognize the magnitude of a fault has lead to 'lost' coal, and to foundation problems for large buildings and bridges. Severe folding, while not common, is important in some regions, and

◁ *Fig. 57* Copies of field sketches of Millstone Grit structures from Bowland, NW England. **A**, Channel deposit of sandstone with a thin coal seam, cut into shale, Claughton Quarries. Note the small accommodation faults at the channel margin. **B**, Sandstone–shale sequence along the Roeburn Valley showing the rapidity with which individual sandstone beds lens out. **C**, Contortion of bedding by slumping in a laminated siltstone. **D**, A small joint-bounded block of marine (Caton) shale showing irregularity of bedding. **E**, Alternations of shale and sandstone deformed into a sharp anticline by valley bulge (see Fig. 61). **F**, Convolute bedding in thin-bedded sandstone.

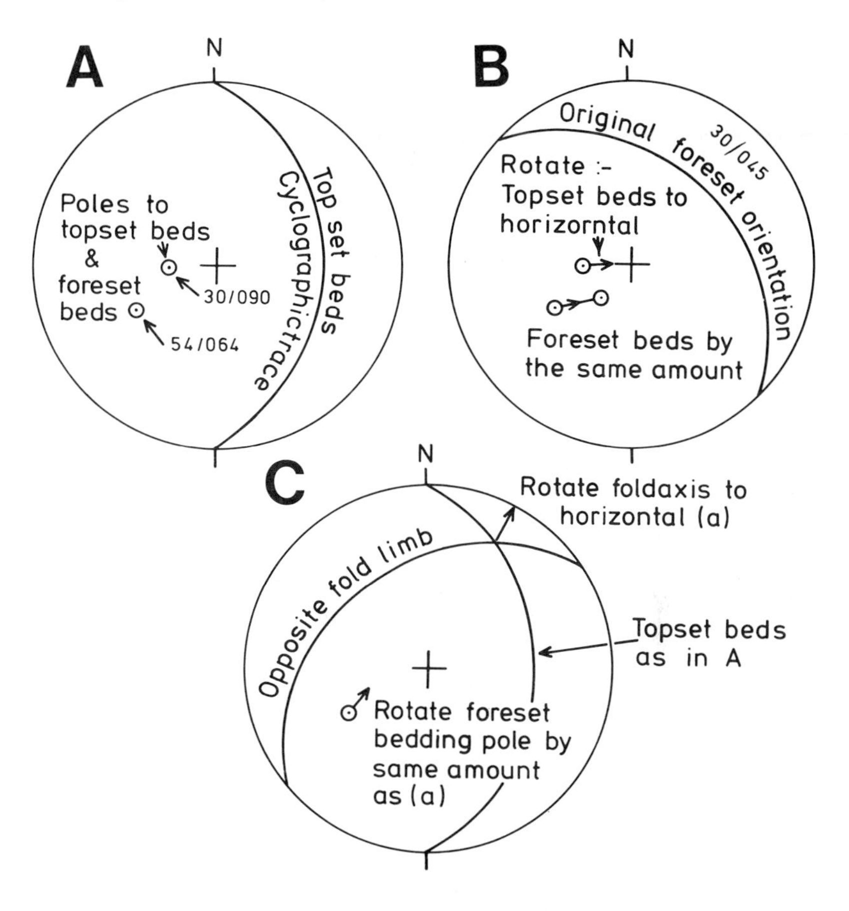

Fig. 58 Deltaic cross-bedding ▷ corrected to its original orientation stereographically by removal of the tectonic dip. The tectonic dip of 30° east (**A**) is rotated to horizontal in **B**, the foreset beds being rotated by the same amount. This method is used where folds have low plunge which is usually the case in the north Pennines. If plunge is appreciable corrections have to be made in two stages. Stage 1 is to rotate the fold axis to horizontal, and stage 2 is as in **A** and **B**.

the Ribblesdale folds and Staffordshire folds ('rearers' because of the highly inclined coal seams) are as complex as many of those in highly tectonized regions (see Fig. 26).

Folding. There are large areas where folding is very gentle with dips less than 10°, but as indicated above there are also regions where folds are steep and unusual although deformation is never intense enough for cleavage to develop. It is important in these cases to take sufficient dip readings so that the three-dimensional geometry of the structures can be ascertained. Without this the true nature of the minor folds shown in Fig. 60 would not be apparent. It is also necessary to beware of that variety of superficial folding known as valley bulge, which can easily be mistaken for a true tectonic structure. Valley bulge is generally restricted to the valley floor and may be well exposed in stream sections, with tight folds usually trending parallel to the valley (Fig. 61). It is caused by inward pressure from the valley sides causing the

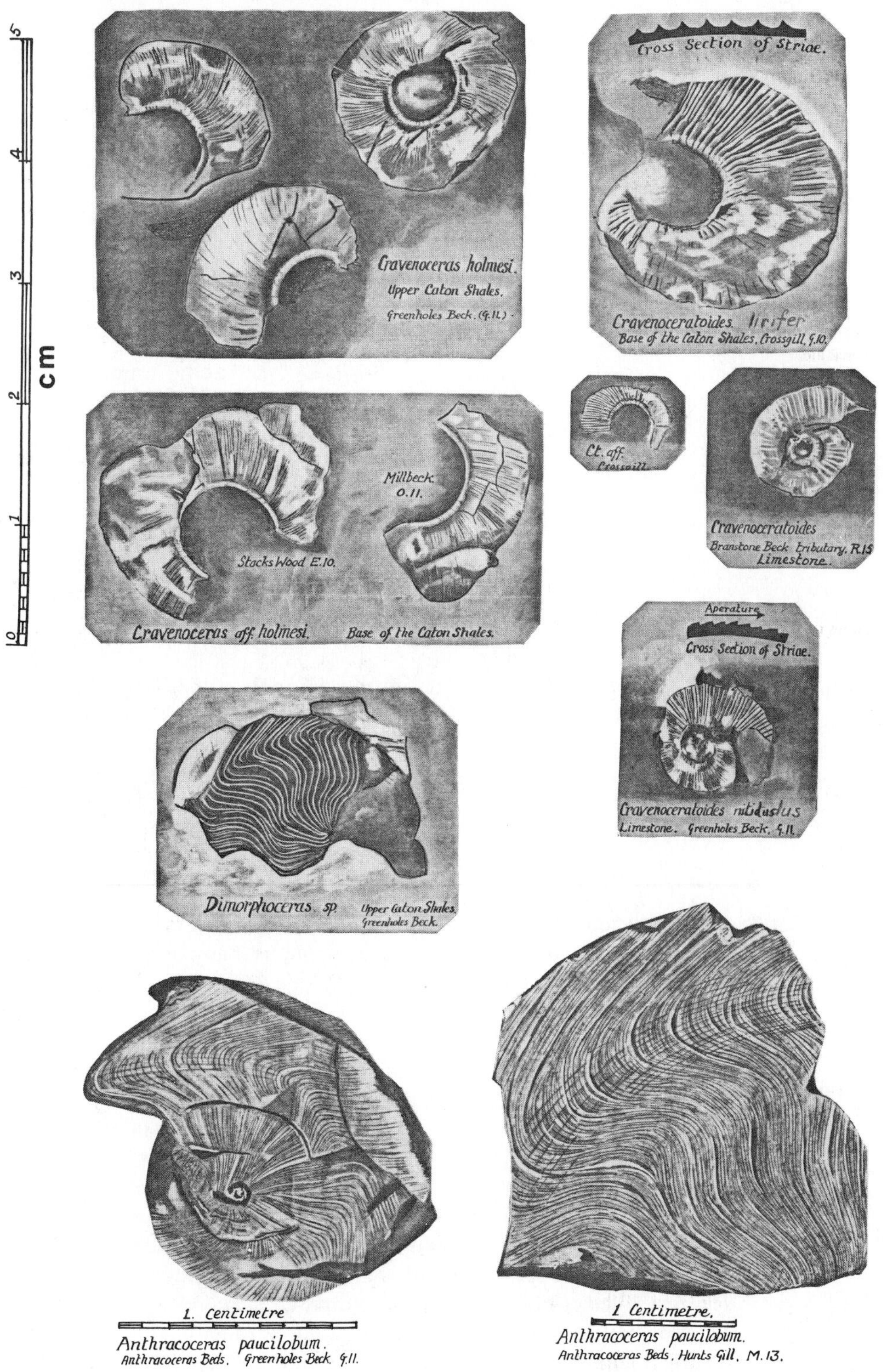

Anthracoceras paucilobum.
Anthracoceras Beds. Greenholes Beck. G.11.

Anthracoceras paucilobum.
Anthracoceras Beds. Hunts Gill. M.13.

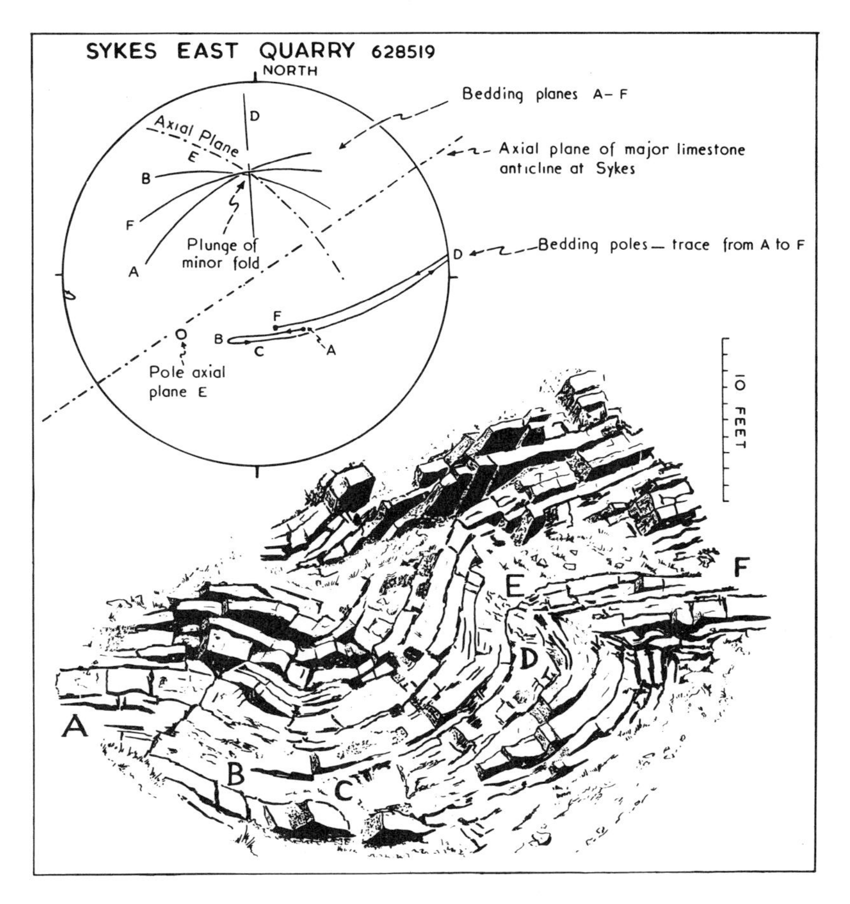

Fig. 60 A method of recording minor folds stereographically. The geometry of the fold is given on the stereographic projection with cyclographic traces and poles A–E corresponding with those on the drawing (Moseley 1962).

◁ *Fig. 59* Drawings of Namurian goniatites (Arnsbergian) from Bowland, NW England prepared for a PhD thesis (Moseley 1954). For thesis work, and especially undergraduate theses, hand lettering (rather than the use of stencils) should generally be adequate. Small crushed fossils in shale are not easily photographed and careful camera lucida drawings may be the best method of illustration.

less competent strata to buckle and can cause problems when engineering works such as reservoirs and dams are contemplated.

Some of the problems encountered in gently folded or tilted strata have already been discussed (page 78), and it has been suggested that a combination of many dip readings and construction of structure contours is the best method for estimating the overall thickness of the sequence, details being added from measured stream and quarry sections.

Faulting. Faulting is more important than folding in the Carboniferous rocks of the Pennines. From Derbyshire northwards to the Scottish borders fault trends are predominantly north–west and north–east, although in some regions there are important northerly and easterly trends. They are mostly normal faults with hades of between 10 and 40° and displacements ranging between 10 and 100 m, but there are a few of much greater magnitude. Students working in these areas

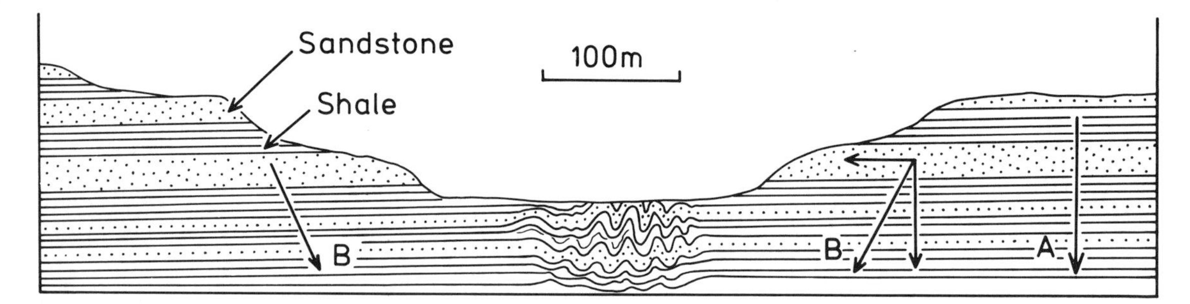

Fig. 61 Idealized section to show the development of valley bulge in horizontal alternations of sandstone and shale. Erosion of the valley results in an inward pressure (B) towards the valley (see Fig. 57E).

should consult the literature and be familiar with these facts and with the theories set forward to explain them.

Most faults, on those rare occasions when they are exposed—usually in stream banks—are seen to form either thin zones of breccia or slickensided clay gouge generally less than 1 m wide. There is no difficulty in understanding why they are easily eroded and infrequently exposed. More detailed investigations will reveal that the bedding generally dips towards the downthrow side of the fault (Fig. 62) and that in addition to the principle plane of displacement, there may be several other smaller ones with parallel trends throwing down in the same direction; including these, the whole zone may be more than 100 m wide.

Calculation of the throw of a fault requires detailed knowledge of the succession, and especially of the faunal marker bands that enable correlation from one area to another. Lithofacies on its own is inadequate for this purpose; there have indeed been many wrong estimations of fault displacement resulting from a mistaken correlation of the 'same' lithofacies across a fault. Fig. 63 gives an example of the way distinctive faunal horizons can be used to estimate fault throw. Mapping across open country also reveals the positions of faults, usually by the terminations of features and escarpments such as those shown on Fig. 52; but, again, estimation of the displacement is dependent on a detailed knowledge of the successions on either side of the fault.

Jointing. In many regions, especially where structures are complex, the joints change in orientation with every change in bedding dip, so that their study, requiring numerous measurements (perhaps many thousands), is time-consuming and in most cases will be beyond the scope of an undergraduate project. A relatively straightforward example of this type of structure is illustrated by Fig. 64. The Carboniferous rocks of the Pennines, however, are not usually severely deformed, and there are circumstances when joint orientation and density (see Doughty 1968) can be studied with profit.

Fig. 62 (opposite) Field sketches from a BSc thesis showing outcrops of small normal faults in stream banks (Artlebeck, Bowland, NW England).

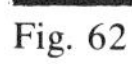

Fig. 62

Fig. 63 A map of part of the Millstone Grit in the Hindburn region, Bowland, NW England. Correlation across faults and the displacement of the faults can be determined from diagnostic faunal bands such as that of *Cravenoceratoides lirifer* which is less than 1 m thick (see also Figs 25, 36 and 52).

Where dip is gentle, joint and fault trends are frequently parallel to each other; this is perhaps to be expected because both are likely to have been formed within the same stress field. Joints are not small or incipient faults, however; they have different origins (Price 1966) and, unlike faults, they are nearly always perpendicular to bedding. Thus although their strike may be the same as that of an adjacent fault, their orientation will be different (Fig. 65). Joint orientation is also controlled to some extent by lithology; for example, sandstone and shale

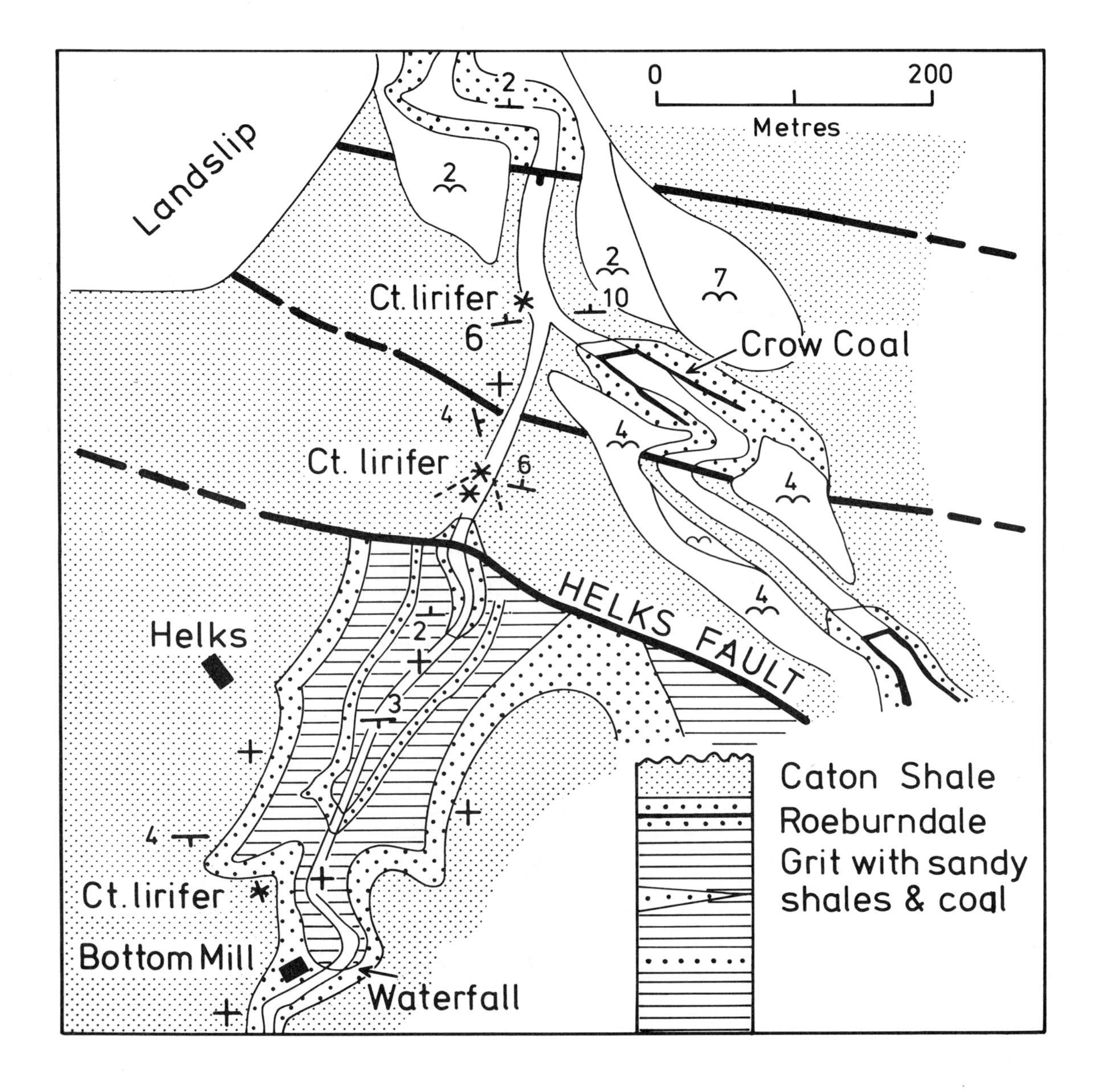

may exhibit different joint trends even when they outcrop at the same locality as shown on Fig. 66, and the surveyor must allow for this possibility when compiling measurements.

With these comments in mind field studies of joints may proceed along the following lines. In Carboniferous Limestone regions there are extensive clint and gryke pavements that represent erosion along master joints and are clearly visible on aerial photographs from which they may be plotted. These measurements need to be supplemented by ground measurements (Figs 47, 48 and 49). In other lithologies joints are not clearly seen on aerial photographs, and interpretations depend on field observation. Sufficient measurements at each locality are necessary to ensure statistical validity (Figs 65 and 66), and it is also important to eliminate bias in deciding which of numerous joint planes should be measured. One method is to select a good exposure and measure every joint within a 2-m radius of a given point, and then move to another position, say 5 m away, and repeat the process. This

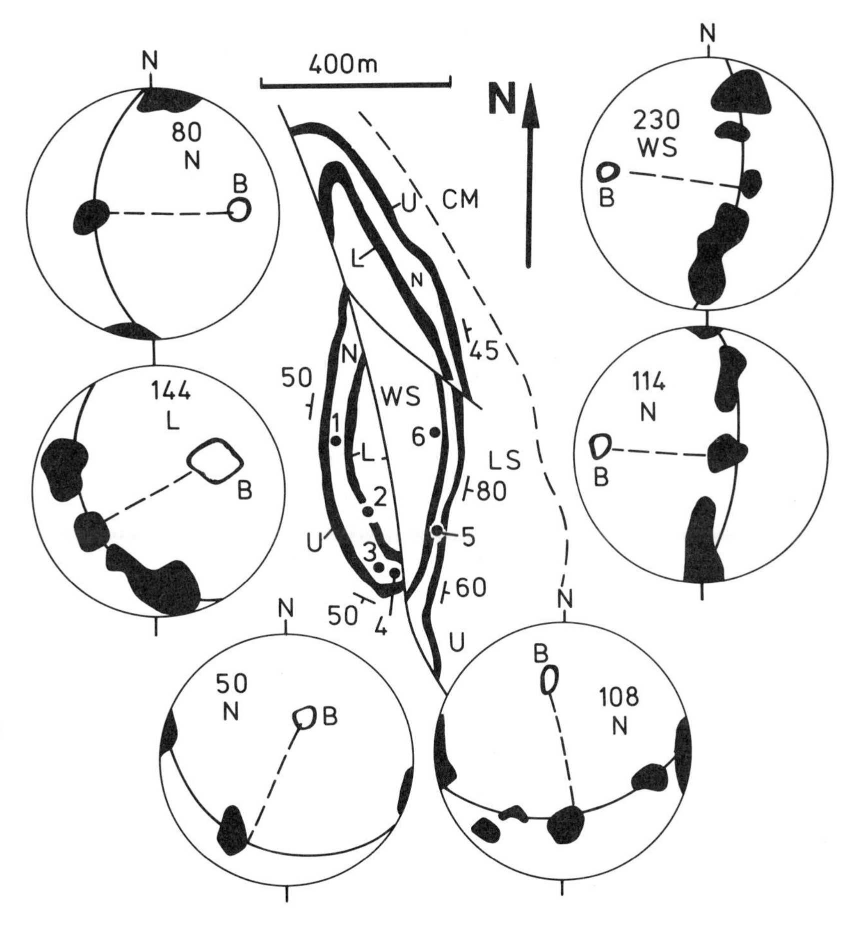

Fig. 64 Measurement of joints on a fold. Note that joint orientation differs according to the position on the fold. 1 and 2 are on the west limb, 3 and 4 in the plunging axial region, and 5 and 6 on the east limb. Had all these joints been plotted on one diagram the stereogram could not have been interpreted. Wrens Nest Pericline in Silurian strata, English Midlands (Moseley and Ahmed 1973).

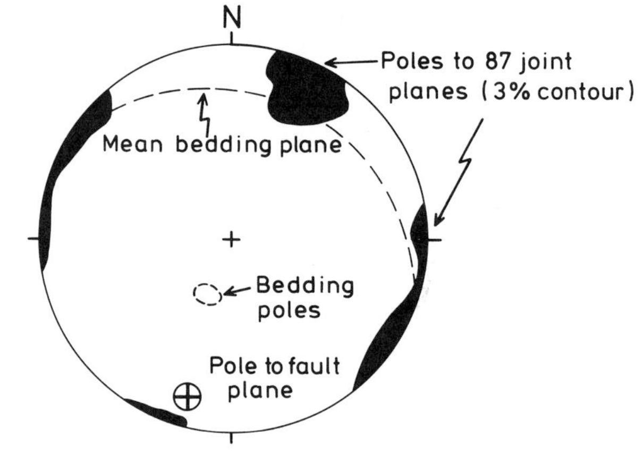

Fig. 65 Comparison between joint and fault orientation across a fault plane (Helks Fault, Fig. 63). Note that there are two joint sets both with different orientations to the fault, although one set does have a parallel trend. Joints of this kind have different origins to the faults (Moseley and Ahmed 1967).

can be done several times until there are 50 to 100 measurements, sufficient for one locality (Figs 49 and 66). In gently dipping strata most of the joints will be near vertical and can be plotted on rose diagrams (but see the comments in chapter 10 (page 61) and Figs 45 and 46). However, if the strata have steep dips it will be necessary to plot the joints stereographically (Figs 64 and 65). Studies of this kind can obviously be applied to rocks of any age, and the reader is directed to papers by Hancock (1969), Hancock and Atiya (1979) and Hancock and Kadhi (1978) for further suggestions.

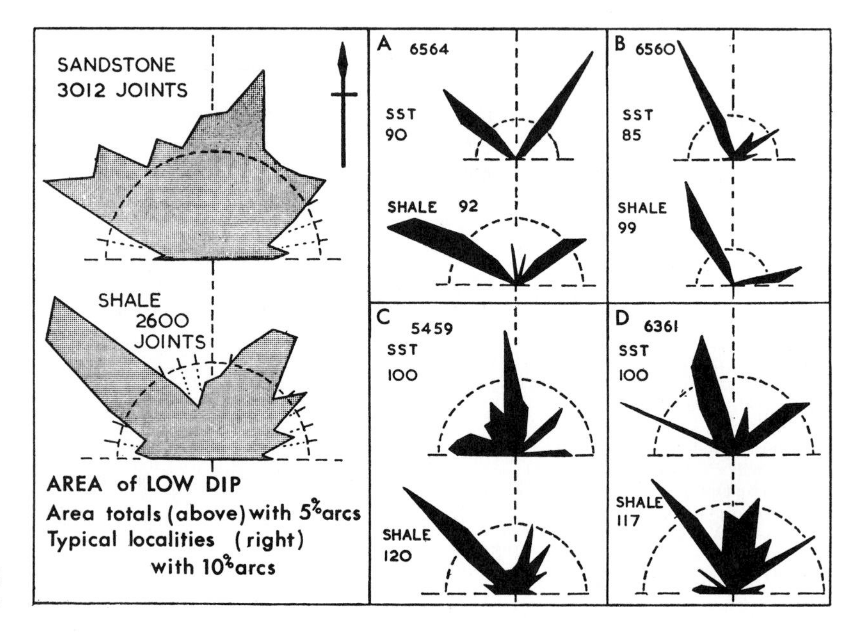

Fig. 66 Joints from the Millstone Grit of Bowland, NW England plotted as rose diagrams (where strata are gently inclined and joints near vertical). Note that trends in sandstone and shale, although at the same locality, can be quite different (A, B, C and D represent four different localities). Also notice that when totals for all localities are plotted together the diagrams become more diffuse, but there are still differences between sandstone and shale (Moseley and Ahmed 1967).

12 Old calc-alkaline volcanic rocks: Borrowdale Volcanic rocks of Ordovician age in the English Lake District

Old volcanic rocks possess all the characteristics and difficulties of interpretation which are to be found in recent volcanics, with the addition of tectonic complications such as the folding, faulting and cleavage imposed in this case by the Caledonian Orogeny (Millward *et al.* 1978; Soper and Moseley 1978). Students and experienced geologists alike can find extreme difficulty both in identifying many of the rocks and in mapping rock units which are not altogether amenable to mapping techniques used successfully on other field problems. It is particularly important when interpreting these older rocks to be aware of the nature of present-day volcanic processes, preferably with first-hand experience.

Calc-alkaline volcanoes are associated with island arcs, active continental margins and subduction, modern examples being seen in New Zealand, the Andes, the Cascade Range of NW America and elsewhere. The 1980 eruption of Mount St Helens in Washington State would have its counterpart in the Borrowdale Volcanics, perhaps in the formation of the Pavey Ark breccia in the Langdales. Older examples are worldwide and range in age from Tertiary to Precambrian; for example, the strongly folded Tertiary volcanics of the coast range of Oregon and Washington, the volcanics of eastern Australia and of parts of Antarctica, as well as other volcanic areas in Britain, such as those of Wales and parts of Scotland.

Difficulties of interpretation of old volcanic rocks are as follows:

1. Initially it is not easy to distinguish between different rock types in the field. Lavas and tuffs can resemble each other; for example, flow-brecciated and flow-banded lavas are easily confused with agglomerate and bedded tuff, and ignimbrite was regarded as rhyolite lava until Oliver's work of 1954 (Figs 73, 74, 77, 78 and 79).

2. A large proportion of the rock outcrops are massive crags cut through by joints with a wide variety of orientations. Tectonic joints are the most common, but there are also cooling joints and joints related to flow structure. The latter will give the dip of the rock but great care is necessary to ensure that meaningful bedding planes are measured. In order to be certain about the dip it is generally necessary

to locate compositional planes, such as alternations of fine and coarse tuffs, or distinctive flow banding of lavas.

3. The Borrowdale Volcanic rocks were erupted as a central vent complex forming part of an island arc flanking Iapetus. It is well known that in recent central vent volcanoes the rock units are usually of small areal extent. Lavas may be little more than streams, perhaps extending some distance in the direction of flow but only a few tens of metres at right angles to the flow; tuffs may be more extensive but can change rapidly from coarse to fine in a few metres, and many of the rocks were erupted subaerially on to eroded surfaces of valleys and hills, which they mantle in a complex way. Since older volcanic areas including the Borrowdale Volcanics are likely to exhibit these features but are also strongly deformed by faulting, folding and cleavage, it has to be decided whether failure to trace (say) a particular lava flow beyond a certain point is because of faulting or simply represents the edge of the flow (Figs 70 and 71). It can be even more difficult to decide between subaqueous and subaerial origins. Some of the bedded tuffs have sedimentary structures which clearly reveal redistribution by water as volcaniclastic sediments, and most of the ignimbrites are likely to have been of subaerial origin; many other rocks remain in doubt (Figs 72 and 75).

It will thus be apparent that those new to the Borrowdale Volcanic rocks will find initial difficulties, and mapping techniques already established may have to be changed. It may be that those with experience of mapping bedded sediments—for example, in adjacent regions such as the Carboniferous Millstone Grit—will now find that it is not the stream courses that give the best information but the rocky crags between the streams. This is because all the rocks are hard and in this respect similar to each other; it is the minor differences brought out by differential weathering of the crags that are important.

Use of aerial photographs

The ground is usually mountainous and open, with good exposure of rock, but it can be extremely difficult to locate accurate positions on maps. It is necessary in these circumstances to use aerial photographs, but they are also needed as an integral part of mapping as indicated in Part I.

An initial photogeological map is prepared (Fig. 7). Aerial photographs frequently reveal escarpments and benches that separate successive escarpments. The former will represent either single lava flows, groups of flows or alternations of different kinds of tuff, and they therefore indicate the original layers of volcanic deposition which, following the Caledonian Earth movements, will now have been folded into various attitudes. The benches between the escarpments indicate similar layers of more easily eroded material such as the amygdaloidal or flow-brecciated tops and bottoms of flows, or interbedded softer tuff bands. Bedding planes can be discerned as dip slopes on the photographs, particularly in tuffs, and the angle of dip can be measured to within 10° using methods given in Lattman and Ray (1965) and in Allum (1966), or alternatively can be estimated empirically once a few field measurements have been taken.

Other important features are gullies, which frequently form straight lineaments seen on aerial photographs to continue from one crag to the next. These lineaments in most cases represent either faults or veins, but there are numerous shorter lineaments which should be recorded; they represent master joints and cleavage (Figs 7 and 67).

A photogeological map will save the surveyor a considerable amount of time in the field—time which otherwise would have to be spent tracing the tops and bottoms of escarpments and the continuity of gully lines. Indeed the significance of many of these features can escape the notice even of experienced surveyors when they are on the ground surrounded at close range by confusing mass crags.

Survey methods

The actual fieldwork should preferably be conducted as follows:

Initial reconnaissance

There should be an initial reconnaissance of the whole region, as far as possible keeping high on the ridges so that good views are obtained. Ground photography, including stereophotography (chapter 8) is important during this phase.

Mapping procedure

Detailed mapping. Detailed mapping should start at a relatively straightforward and easy part of an area, generally determined from aerial photographs. It is bad for morale if at first nothing makes sense, and this may happen if a start is made in an area which is complex.

Reference to point of access. It may be that there will be only one easy point of access to a mapping area—perhaps a road running along one side, with the ground rising steeply into mountainous areas beyond the road. Much of the fieldwork may be 2 or 3 hours' walk and 500 m above the road, so it is important to plan the area to be covered each day in such a way that one rarely has to spend valuable time simply walking across areas that have already been mapped. The ground can be covered systematically in the manner shown on Fig. 68; a photogeological map helps (Fig. 67), since one is not tempted to follow lineaments from end to end in the field when they can be satisfactorily plotted from the photographs.

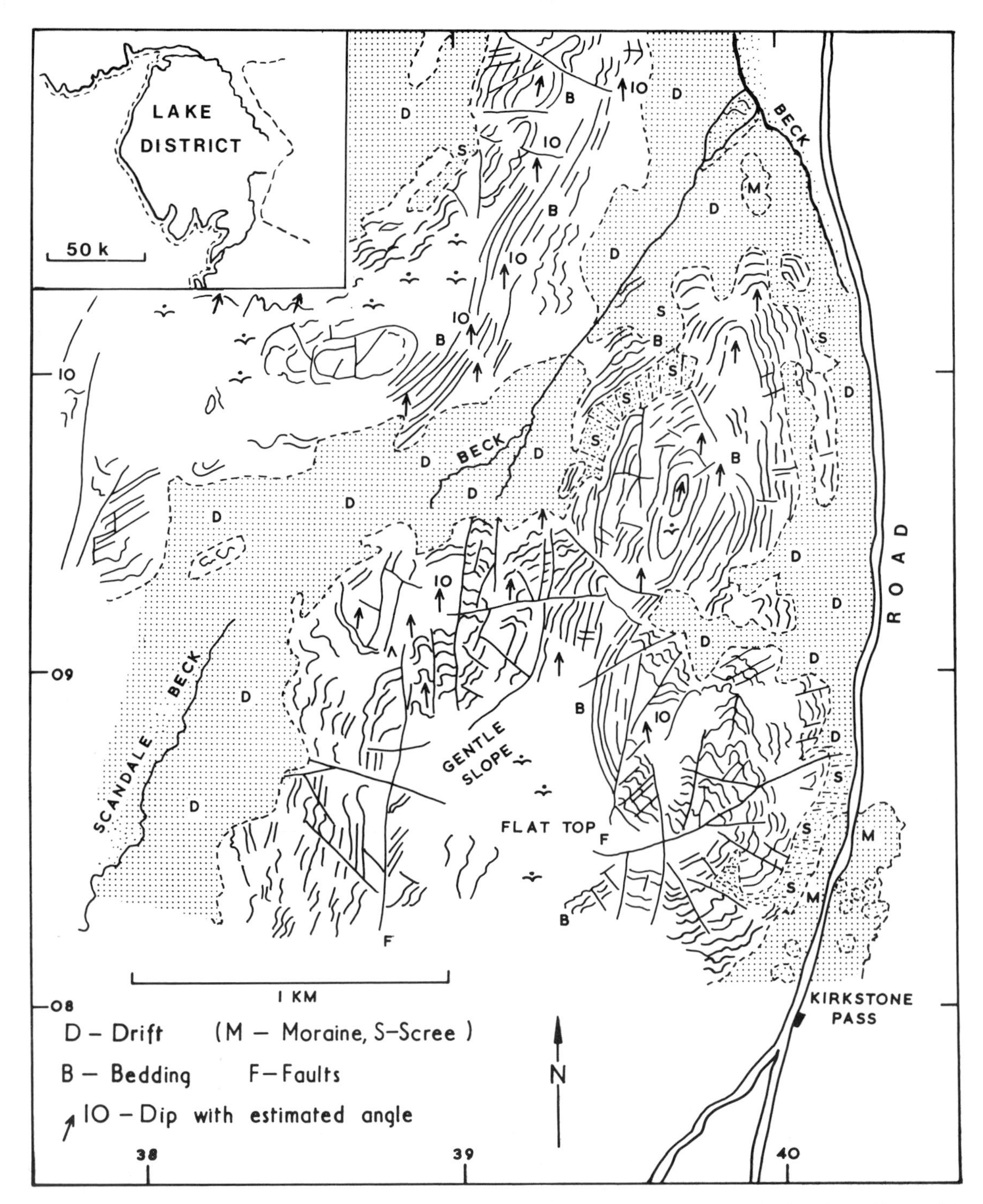

Fig. 67 Photogeological map of the Kirkstone area of the English Lake District. Most of the legend for Fig. 7 applies to this map. Dip arrows rather than the strike symbol are used because on photogeological maps there are many short lineaments (joints, cleavages, etc.) which can be confused with the latter (see Fig. 12). It is usually possible to plot fairly reliable drift boundaries and sometimes to interpret their nature, but it is dangerous to rely too heavily on the photo interpretation, especially for lithologies. Subsequent field survey revealed that most of the rocks are bedded tuffs, but there are flow-banded dacite flows at 387 101 (Little Hart Crag) and 392 107 (within the enclosed lineaments), and an andesite flow 500 m NW of Kirkstone Pass. The terrain is mountainous (see Fig. 68). The inset map shows the location of the Lake District in northern England.

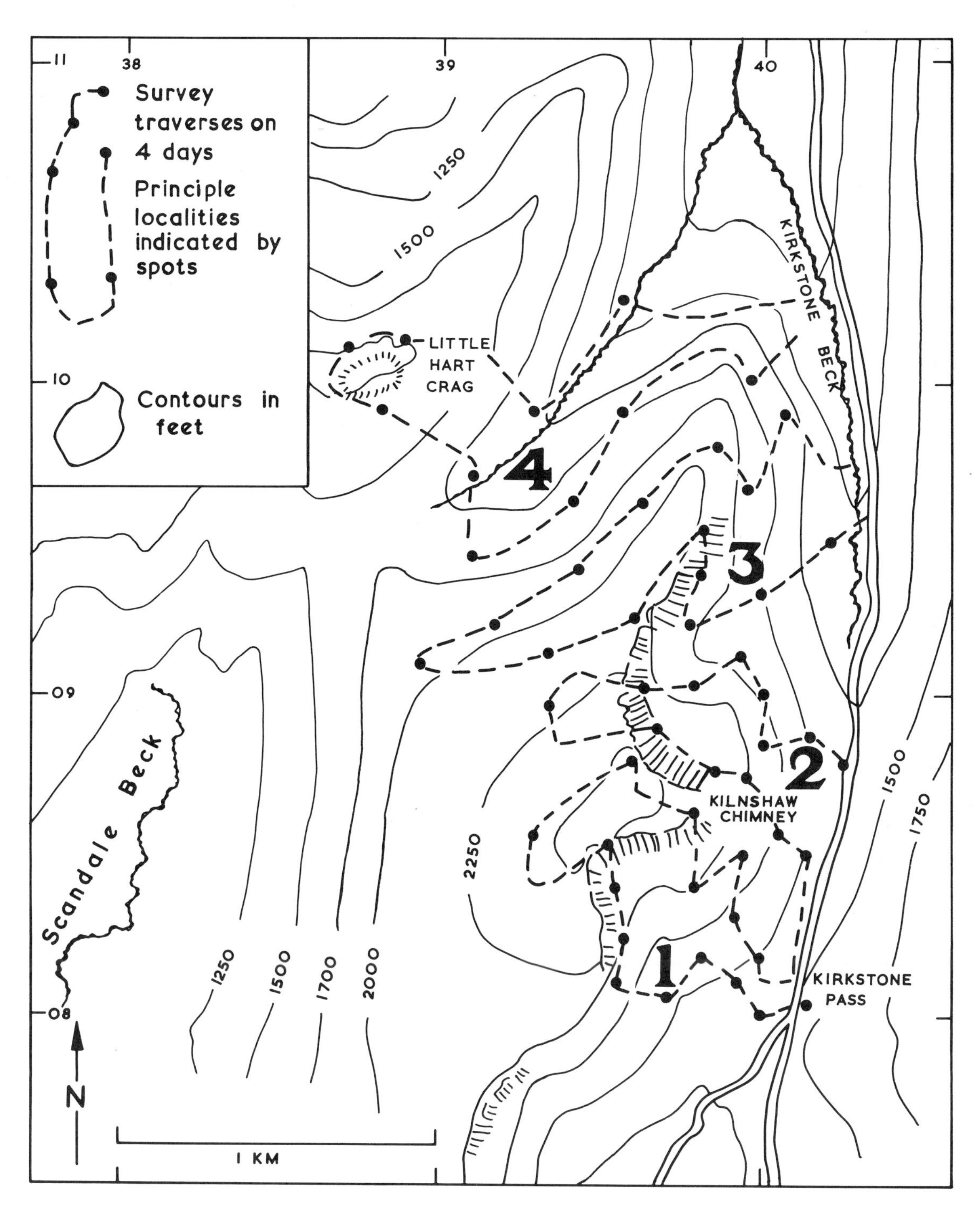

Fig. 68 Planned survey traverses for four days across the area of Fig. 67. A photogeological map makes it much easier to plan approximate field traverses in advance. The traverses are designed to avoid unnecessary walking and climbing across rough ground already surveyed. In this case all start and end at accessible points along a road. It will be understood that planned traverses are provisional only, since, as fieldwork progresses some areas may require more and others less time than the original estimates.

Use of aerial photographs in the field

Since it is often the case that a topographical map will have insufficient information for establishing an accurate position, it will be necessary to use aerial photographs for recording field data. This can be done with far greater ease and reliability if the aerial photographs are placed in a map case for stereo-viewing (chapter 4) and a pocket stereoscope carried. Ideally two sets of photographs should be carried, one for reference and the other for recording geology. Financially this is beyond the reach of most students, and a compromise is to carry photocopies of the photographs. It has to be stressed that no permanent marks should be made on the reference set of photographs since these will obscure the features which they mean to illustrate. Photocopies can thus be used as field maps, an alternative being to use transparent overlays on the photographs. The original photogeological map will already have too much information recorded on it to permit further additions but it should nevertheless be carried in the field for reference. The field map (photocopy or overlay) can be used to plot structural measurements, rock types, localities referred to in field notes, and outcrops of boundaries particularly where they differ from those of the photogeological map.

Fig. 69 (opposite) A view SE from Green Gable in the English Lake District with the Langdale Pikes (left) and Bowfell (right) visible in the distance. This photograph is part of a stereopanorama, and illustrates well the advantages of viewing a mountainside from a distant vantage point, both before and during a survey. There are many parts of this area where the geological structure is easily seen from a distance, but becomes confused in a jumbled mass of joint-infested rocks when one stands amongst the crags. The interpretation of the photograph follows Oliver (1961), and was drawn from a photocopy of the original, using white and black inks. Diagrams of this kind can aim to look realistic and artistic, as with Fig. 70, or can be purely functional, as in this case.

The distance from Sty Head Tarn to Allen Crags is 2 km, and their difference in height is 346 m. A, Airy's Bridge Ignimbrite; T, Seathwaite Fells Bedded Tuff (with 'Th' a fine-grained flinty variety); L, Lincomb Tarn Ignimbrite; E, Esk Pike Hornstone (bedded tuff); F, faults.

Field observations

The sequence of field observations during rock identification would generally be as follows:

1. A general impression is obtained of the whole outcrop seen from a distance. Lavas are often more massive, with less layering than tuffs, and the latter also take a stronger cleavage. Prominent gullies (faults, veins or master joints) show up clearly (Figs 69, 71 and 76).

2. At close quarters careful inspection of weathered rock surfaces is essential. They are generally characteristic of the rock types, but at times freshly broken specimens and even thin sections can be misleading. For example, flow brecciated lava can be mistaken for coarse tuff, but in the particular environment of the English Lake District the matrix is generally harder than the angular fragments and stands out on weathering, with the fragments recessed. The opposite is the case with coarse tuffs in which it is the fragments that stand out on the weathered surfaces. The detailed nature of the flow breccia matrix also shows up on weathered surfaces and can frequently be seen to be of igneous origin rather than the fragmental origin of a pyroclastic rock. Bedded tuffs and flow banding of lavas are also quite characteristic on weathered surfaces, the former closely banded in alternations of fine and medium grades, often with complex sedimentary structures, and the latter with a more gradual change from one layer to another. One other texture commonly seen is the eutaxitic texture of ignimbrite

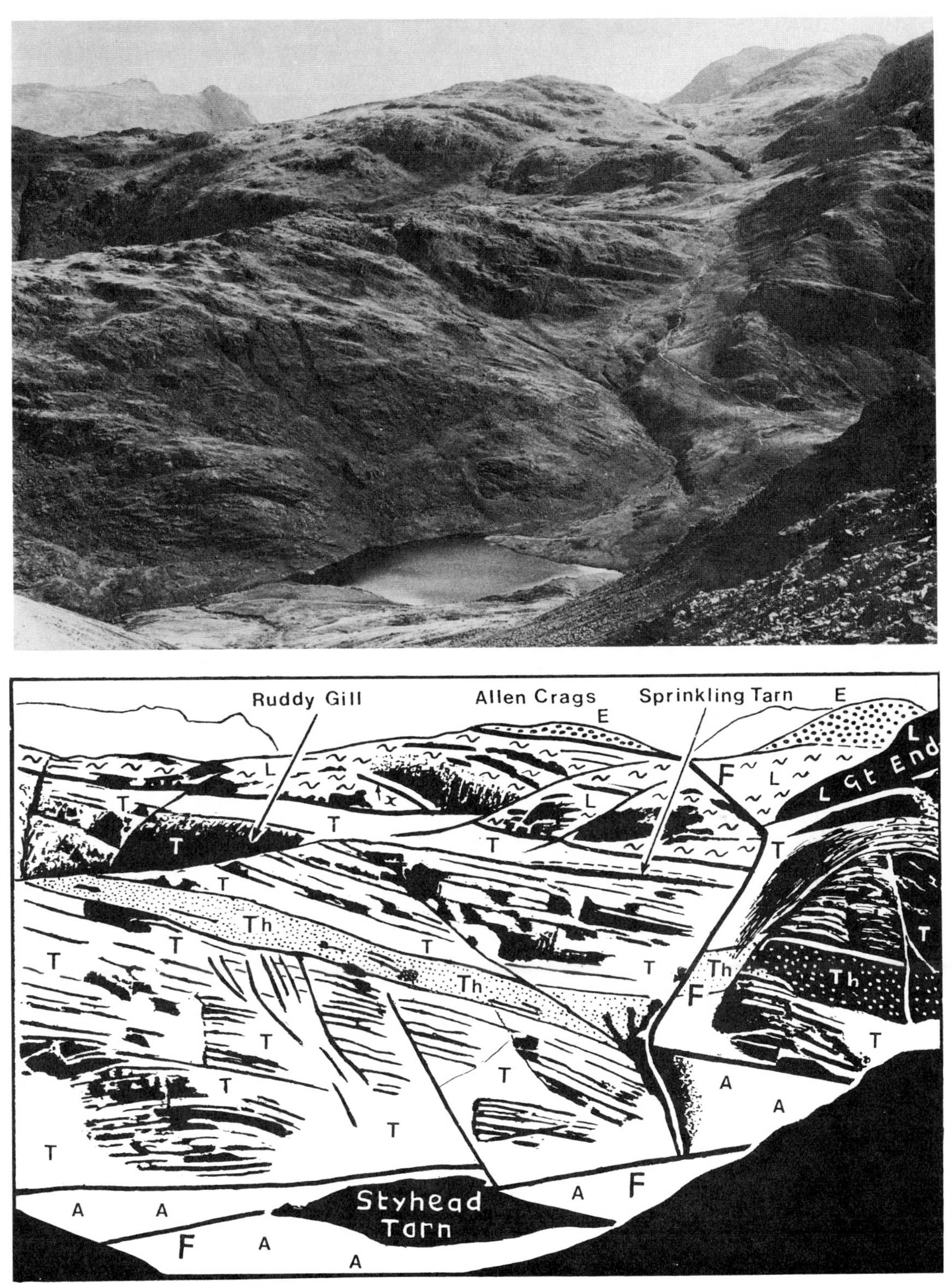
Ruddy Gill
Allen Crags
Sprinkling Tarn
E
E
L
L
L
F
L Gt End
T
x
Th
Th
Th
F
A
F
Styhead Tarn
F

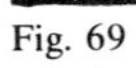
Fig. 69

generally weathering to dark chloritic streaks in a pale acidic rock (Figs 72 to 75, and 79).

3. The rock is then broken and the fresh surface inspected with a hand lens; this may reveal some additional facet not seen on the weathered surface (identity of a phenocryst or vesicular infilling, etc.). A small or a large specimen is then collected. The size of the specimen will be decided in relation to subsequent laboratory study as indicated below.

Fig. 70 A drawing of the view east to Barton Fell, Ullswater, English Lake District. The massive escarpments are mostly basaltic andesite lava (Moseley 1960), and the separating benches are formed from bedded tuffs. When lavas suddenly terminate, as shown in the drawing, several explanations may be possible. (1) An oblique fault may cut across the outcrop displacing hard lava against softer rock. (2) The termination may be the side or end of the lava flow. Modern lava flows show that such terminations can be abrupt. (3) There may be an unconformity, with overstep of the lava bands. In this particular example it will be noticed that several escarpments (lavas) end in turn against one line at the bottom of the fell, and the volcanics are faulted against softer, mostly unexposed Skiddaw Slates that occupy the foreground.

Specimen collection

During mapping it is best to collect small specimens at each locality, rather than to rely entirely on field identification. Larger specimens for other purposes (geochemistry etc.) can be collected periodically as required. This will result in numerous small specimens (perhaps 1000 in a field season), identification of which can be confirmed later at base.

The next stage is reserved for 'base camp' in the evening, when the day's specimens can be examined more carefully, preferably using a binocular microscope; the final stages will await laboratory investigations.

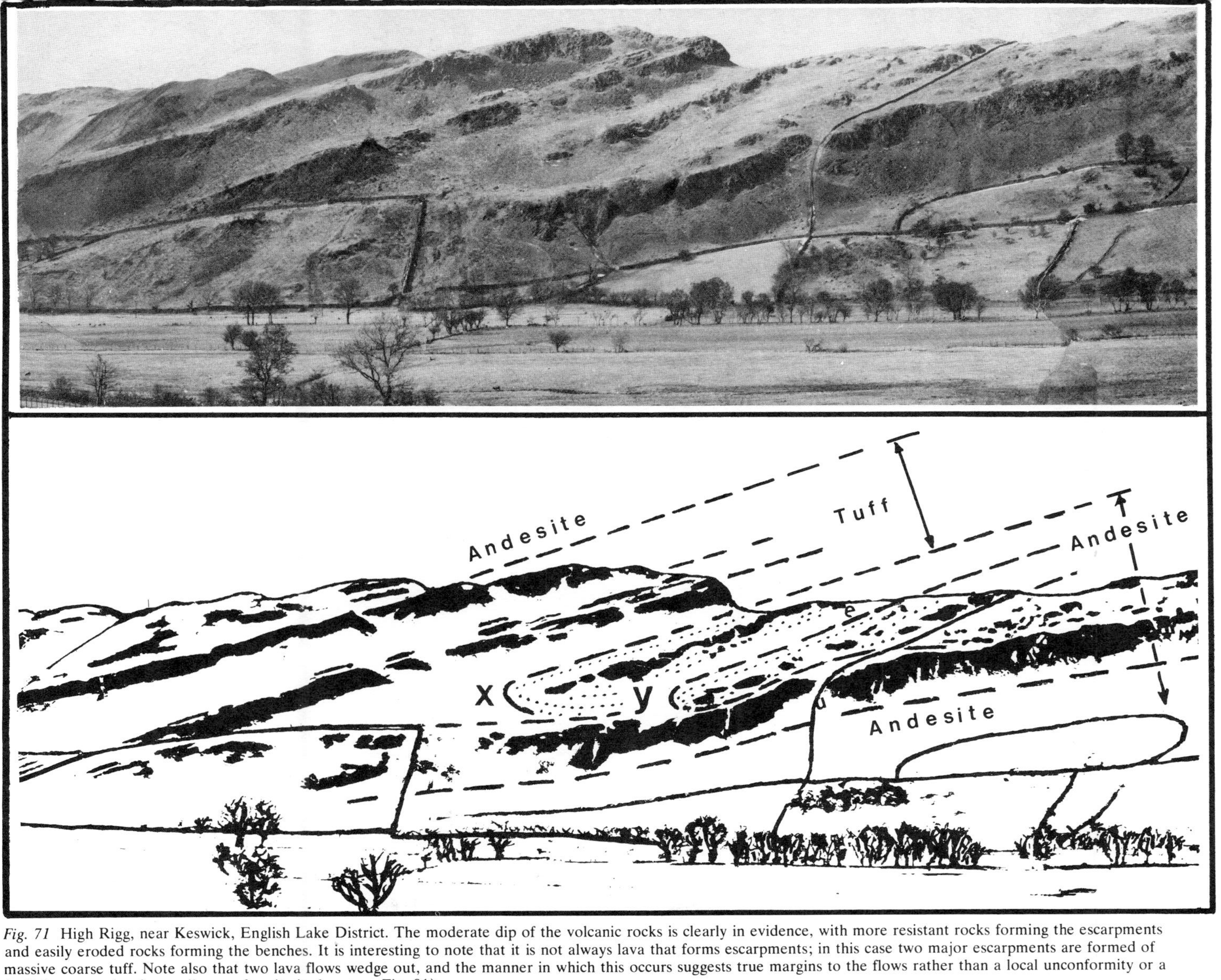

Fig. 71 High Rigg, near Keswick, English Lake District. The moderate dip of the volcanic rocks is clearly in evidence, with more resistant rocks forming the escarpments and easily eroded rocks forming the benches. It is interesting to note that it is not always lava that forms escarpments; in this case two major escarpments are formed of massive coarse tuff. Note also that two lava flows wedge out, and the manner in which this occurs suggests true margins to the flows rather than a local unconformity or a fault (an example of a small unconformity is shown on Fig. 81).

Fig. 72 Hopper Quarry, Honister, English Lake District. The outcrop of the Hopper 'Vein' is worked for ornamental slate, and on this occasion was visited by a Yorkshire Geological Society party. The main part of the quarry is in bedded tuff (see Fig. 75) with the bedding cut by a steeply inclined cleavage. An adequate survey of this quarry would require a full day. (Photograph by J. Denys Hind, Cockermouth.)

Fig. 73 Stereopair of Kevin Smith but also showing bedded tuff cut into by an andesite lava flow (High Stile, Buttermere, English Lake District). Closer views of outcrops are the next stage of an investigation following the distant views illustrated by Figs 69, 70 and 71.

Fig. 74 Flow-brecciated lava is easily confused with coarse tuff and agglomerate in the field. Weathering conditions in Britain usually result in fragments being recessed compared with a more resistant matrix in flow breccia (above), whereas the reverse is true for coarse tuff (below).

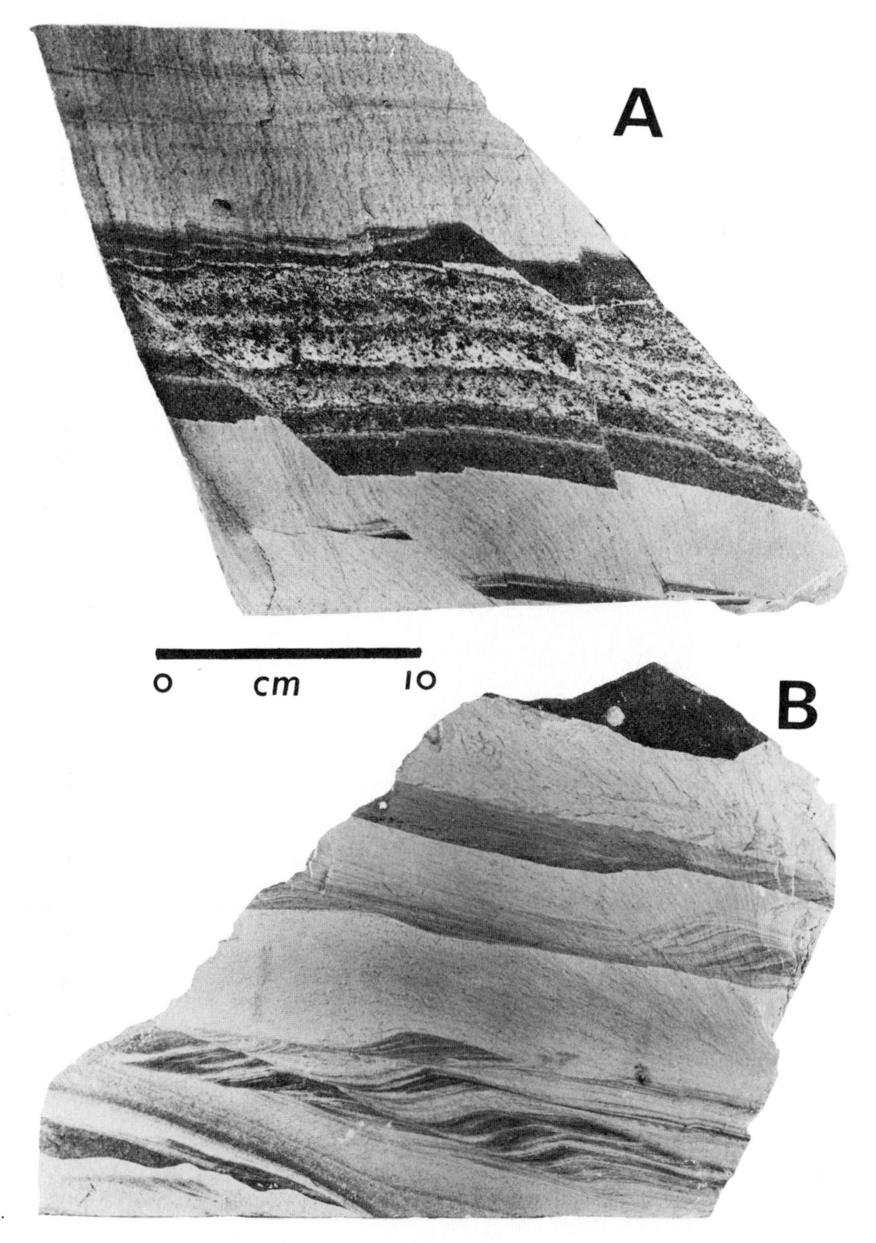

Fig. 75 Bedded tuff, Hodge Close, Coniston, English Lake District. This outcrop has been much worked as an ornamental stone (see also Fig. 72). It is a subaqueous volcaniclastic sediment. A survey project that included rocks of this type would be expected to record the variety of sedimentary structure, and it would be necessary to collect a large number of specimens. In **A** fine and medium grained tuffs are affected by small 'accommodation faults' and in **B** similar tuffs show well-developed current lamination.

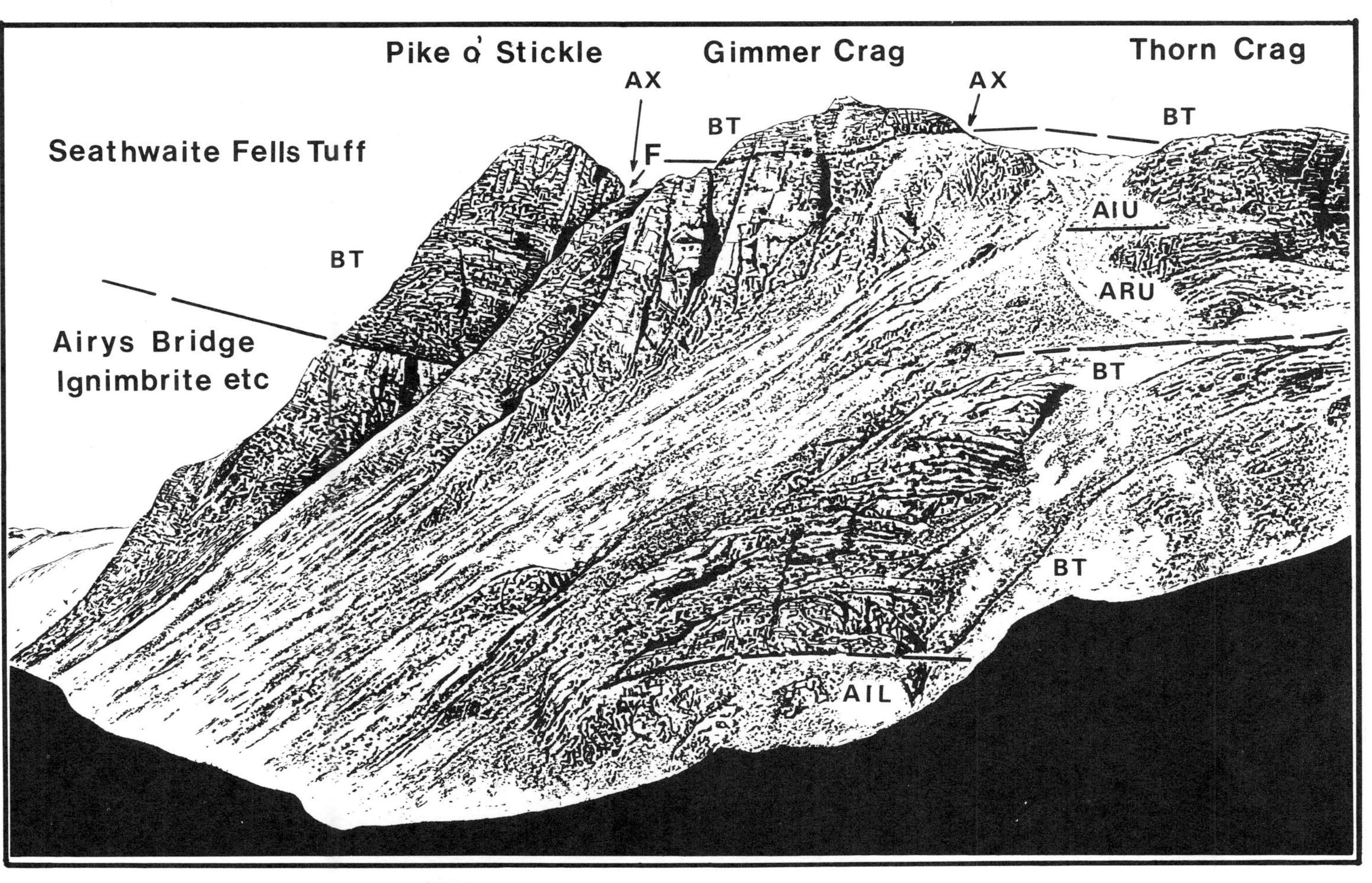

Fig. 76 A drawing of the Langdale Pikes in the English Lake District from the south. The vertical height from the lowest exposures (AIL) to the top of Pike O'Stickle is 500 m, and the volcanic sequence is: AIL (lower Airy's Bridge Ignimbrite), BT (bedded tuffs with coarse tuffs), ARU (upper Airy's Bridge Rhyolite (dacite)), AIU (upper Airy's Bridge Ignimbrite), BT (bedded tuffs of the Seathwaite Fells Tuffs, see Fig. 69). AX shows the location of two Neolithic stone axe factories, which worked the lowest few metres of the Seathwaite Fells Tuff. Note that the ignimbrite is a more massive, harder rock than the bedded tuffs and forms steep cliffs, in particular the well-known rock climbing precipice of Gimmer Crag. Figs 76 to 79 show the stages of investigation starting with a distant view and ending with close-up investigation of the rock.

Large-scale maps

It is often the case that an outcrop will contain too much interesting detail for all to be plotted on an aerial photograph or map, in which case enlargement maps are required (Fig. 81) sufficient to permit the plotting of these extra data (compare Figs 80 to 82). This information will normally form a part of field notes and will facilitate plotting additional dips (cleavage, bedding, master joints). If still more structural and volcanological data are necessary, they are best recorded under locality numbers in additional field notes.

Geological boundaries and other 'mapping lines'

One of the objects of a detailed survey will be to map as many of the volcanic units as possible, for example, down to single lava flows. The mapping lines, however, do not always conform to the expected. Erosion may pick out junctions between vesicular lava tops and massive central parts rather than tops and bottoms of flows, and on other occasions several flows may be welded together so that they map as one unit. Similarly ignimbrites may be subdivided into both cooling and flow units in some localities, but this can be obscure in others. In its final form a field map should record and explain the meaning of the 'bedding' features already plotted photogeologically. Many of these will turn out to be little more than ledges in uniform lithologies, generally where there are minor changes—for example from bedded to unbedded tuff—but these mapping lines should not be dispensed with since they clearly reveal outcrop patterns and these in turn emphasize geological structures (Figs 7 and 67).

Fig. 77 The Yewdale Ignimbrite, Coniston, English Lake District (Millward 1980). A large fallen block of columnar jointed ignimbrite, which, on close examination, exhibits a eutaxitic texture similar to the specimen of Fig. 79.

Fig. 78 Nodular ignimbrite on Kidsty Pike summit, near Haweswater, English Lake District. The nodules are believed to be formed by late-stage vapour phase activity (Millward 1980). ▷

Fig. 79 Ignimbrite from Knock Pike, east Cumbria, Cross Fell Inlier. There is a well-developed eutaxitic texture, which is common in ignimbrites and can make field identification easy. ▽

Base-camp work

On return to base in the evening it is essential to go over the day's work. All the relevant data plotted on aerial photographs (photocopies or overlays) and those from photogeological maps should now be transferred to a final 'field map' (1:10 000 or larger scale). It is important (1) to get the maximum information on to the field map, (2) to ensure that this information is self-explanatory and not simply a mass of numbers (e.g. locality numbers), (3) at the same time not to go too far and clutter the map to the extent that it becomes illegible. A balance has to be achieved between the information recorded on a map and that which is in field notes.

Ground photography

Ground photographs, many of them stereopairs, will be important and should be taken along the lines indicated in chapter 8.

In conclusion it can be restated that inexperienced surveyors almost invariably find difficulty during the first few weeks of mapping on this type of terrain. The procedure outlined here should help to overcome some of the difficulties. It should finally be remembered that the aims of a survey are likely to be not only the elucidation of rock succession and structure, but also an understanding of vulcanicity, and in this latter respect it is valuable to have had opportunities to study areas of recent calc-alkaline volcanic activity, such as the Cascades of north-west America. Complexes of this type, ancient or modern, will contain numerous diachronous and cross-cutting units as well as the parallel bedding of some lavas and tuffs. Vents consisting of coarse pyroclastics are likely to be common, and there will be igneous intrusions for which age dating may be necessary before age relations with the other rocks can be established. The St John's Microgranite is an example of this kind (Wadge *et al.* 1974).

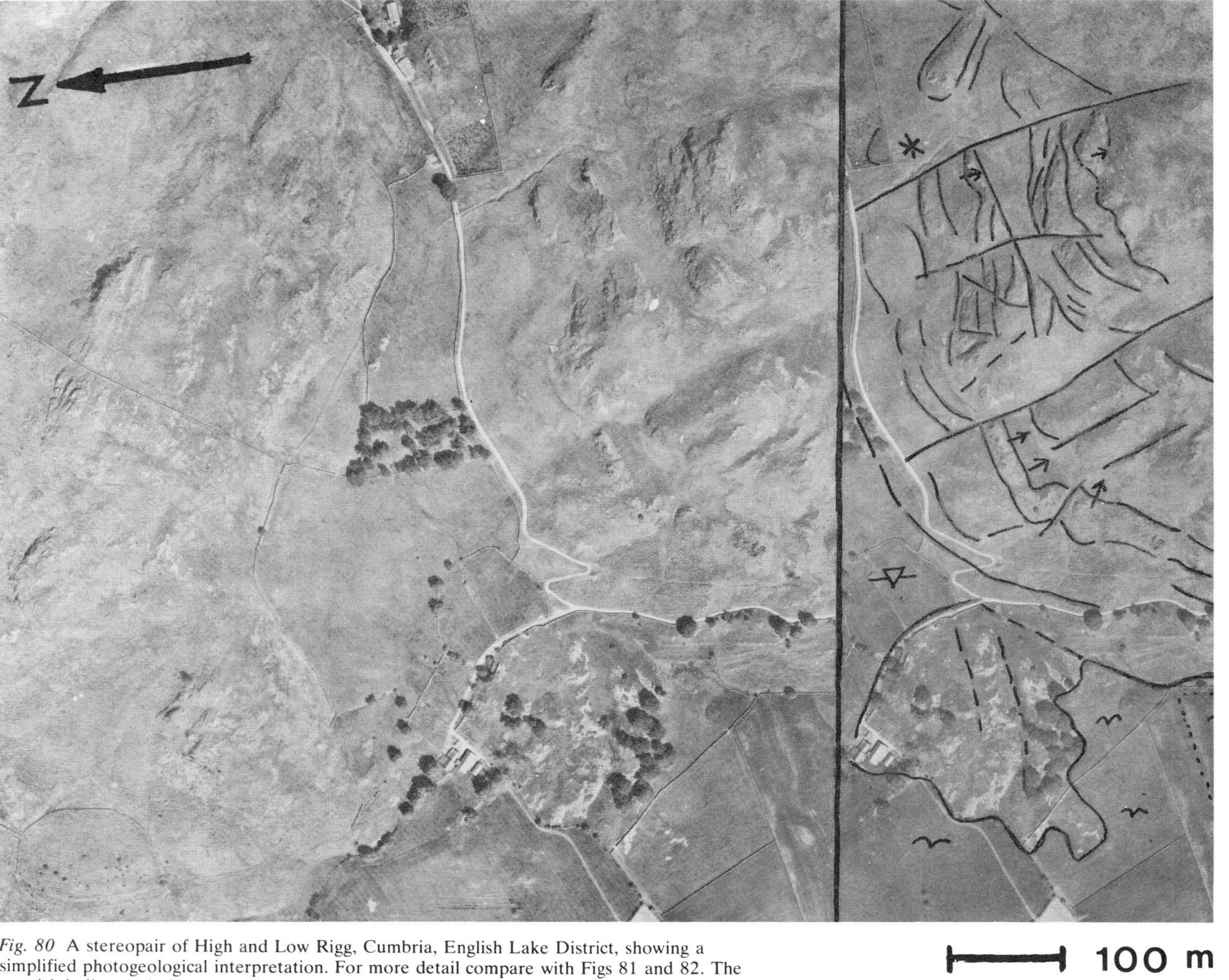

Fig. 80 A stereopair of High and Low Rigg, Cumbria, English Lake District, showing a simplified photogeological interpretation. For more detail compare with Figs 81 and 82. The asterisk indicates the same position on each figure. (Reproduced from the Ordnance Survey aerial photograph with the permission of the Controller of Her Majesty's Stationery Office. Crown Copyright reserved.)

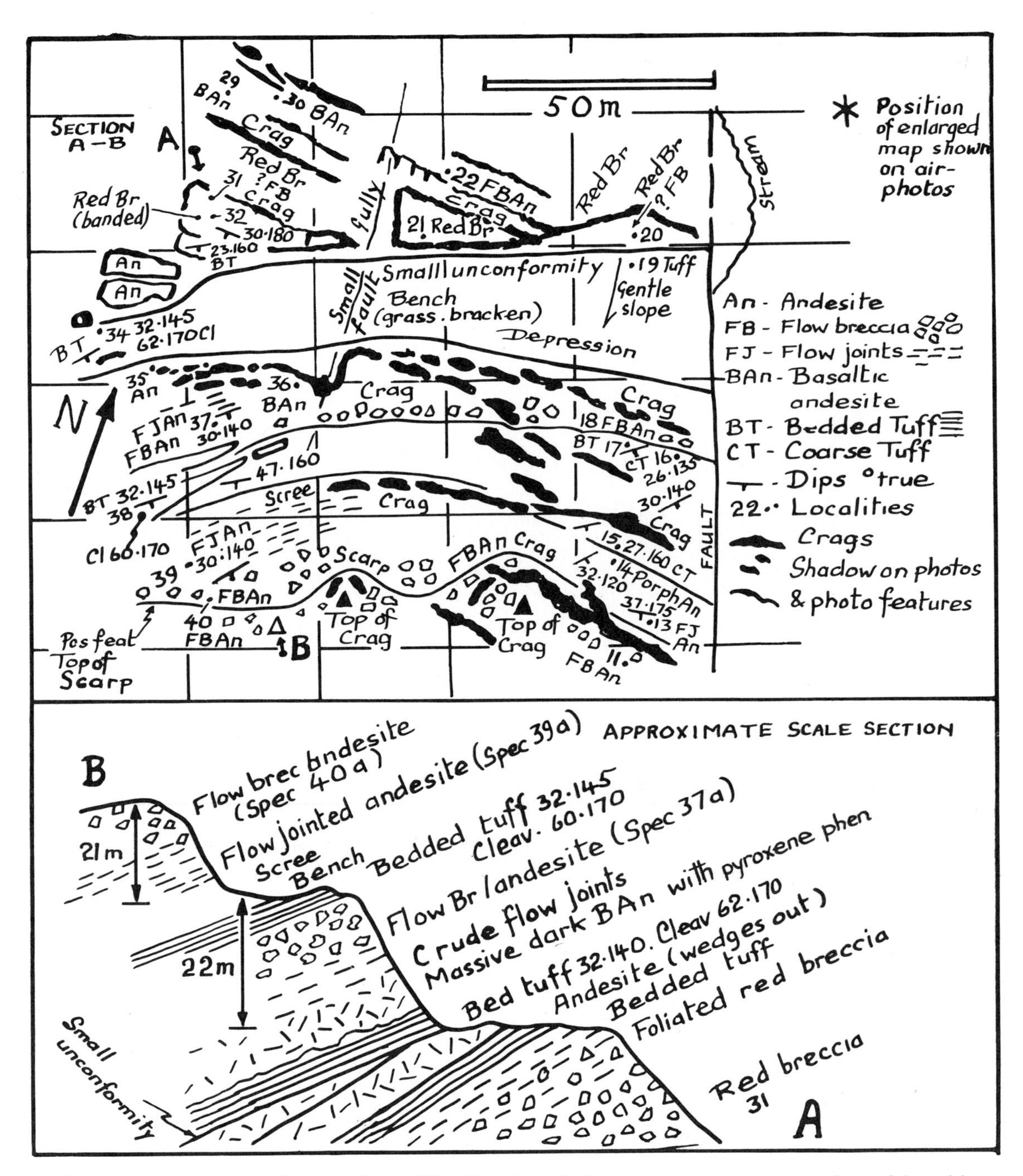

Fig. 81 A copy of an enlargement field map of part of High Rigg, Cumbria. The map is a sketch enlargement of part of the aerial photograph (Fig. 80) and was constructed by drawing small squares on the photograph, and transferring photographic detail to larger squares in field notes (see also Fig. 27). Use of different colours on the original permitted plotting of more detail than is shown on this diagram.

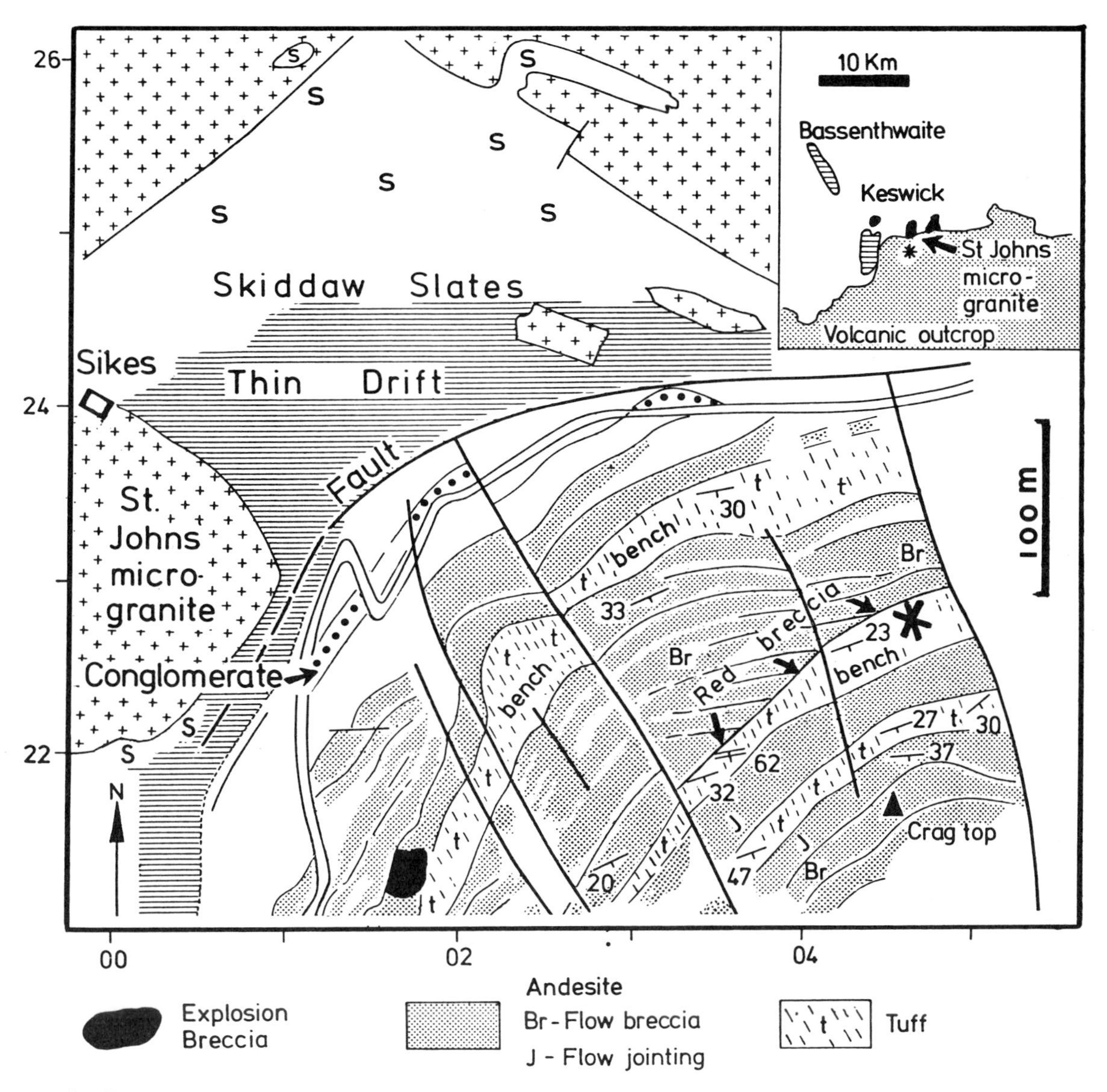

Fig. 82 A published map of High and Low Rigg, Cumbria (Moseley 1977), the earlier field stages of preparation being shown on Figs 80 and 81. The Borrowdale Volcanics are here separated from the Skiddaw Slates and St John's Microgranites by a fault. The latter is intrusive into Skiddaw Slates and is known to be the same age as the last stages of the Borrowdale Volcanics (Wadge *et al*. 1974).

13 Quaternary deposits of the last glaciation in the Cheshire Plain, English Midlands

A geological survey of a region consisting entirely of unconsolidated glacial, interglacial and postglacial deposits requires a completely different approach from that used when mapping solid rocks, and this is especially so in low-lying flat terrains. A first impression of the type of country referred to in this example may be one of a flat plain completely unexposed and without geological features. Analogous regions in other parts of the world are commonplace, especially in those parts of the Northern Hemisphere which were covered by Quaternary glaciers. Apart from other parts of Britain, the plains of north Germany and Poland, and much of southern Canada and the northern United States exhibit similar features.

In Cheshire practically all the ground is farmed, generally mixed arable and pastoral, and is divided into fields, usually by a combination of hawthorn hedge and wire fence. When students are introduced to fieldwork of this kind they scarcely know where to start or what to do and, longing for the mountains, are inclined to retreat to the nearest inn muttering about the importance of hard-rock geology. However, geological surveyors are usually required to be versatile, and this variety of geology is so important to civil engineering that it should be part of every curriculum. In fact it becomes more interesting the more one does, and many are happy to take up full-time employment in 'drift' mapping. In Cheshire there are clear advantages in the networks of roads, and in the hedges, fences and farm buildings, all accurately portrayed on 1:2500 and 1:10 000 (or the existing 6 inches to the mile) maps, so that providing the farmers are treated with respect there is no difficulty of access to the ground, or in plotting exact locations. On the other hand there are disadvantages in the absence of exposure. Clay pits and sand pits are rare, road cuttings, sewer (and other) trenches have to be examined within days, otherwise they are filled in and grassed over. It is these artificial sections which provide the best evidence for interpretation of the glacial environment and modes of deposition, and exposed sections should be carefully measured, photographed and specimens collected without delay. For the remainder it is necessary to 'make exposures' using augers, a relatively easy task in these unconsolidated deposits, and should it be important to expose a more continuous section at a particular locality, it is sometimes possible to hire a trench digger for the purpose if funds are available (if not, borrow a spade). The drift deposits are likely to include alluvium, peat, solifluction deposits, various types of till, lacustrine clays and sand and gravel, and they will have been formed during glacial, interglacial and postglacial periods. Generally they will occur as lenticular deposits, especially in morainic areas, where one cannot rely on individual beds persisting for more than a few yards. In regions away from moraine, however, there are individual drift sheets, till sheets in particular, which can be traced for several miles. The complexities of field relations indicated above give rise to considerable difficulty in the correlation of adjacent sequences, and it is mistaken (although tempting) to equate sand and clay sequences from adjacent sections on the basis of lithological comparability. Indeed the problems of correlation can be compared with those experienced in some volcanic terrains, rather than with those of normal stratigraphical successions; it is inviting trouble unthinkingly to use methods which could be used successfully in the mapping of, for example, Carboniferous sediments.

Glacial deposits in a particular area are quite likely to represent several glaciations, but this may be difficult to establish, and ultimately will depend on the recognition of interglacial organic deposits with diagnostic fauna and flora. The most recent of the glacial episodes (Weichselian or Devensian) is strongly represented in regions of 'newer drift', and there are thick deposits belonging to this phase covering the greater part of the Cheshire Plain, generally recognizable by the 'depositional' rather than 'erosional' topography. This type of topography is dominated by enclosed hollows and irregular mounds and not by eroded valleys separated by interfluves. It has to be emphasized that studies of present-day glaciers and their deposits help enormously in explaining older glacial deposits, indeed without such studies correct interpretation of the latter would be virtually impossible (Figs 131 and 132).

With these background points in mind the following procedures are suggested when mapping an area such as Cheshire.

Use of aerial photographs

Drift features

Drift features can be plotted and a photomorphological map prepared (see below).

Deposits

Some of the deposits can be provisionally determined. Clay, sand and peat can usually be distinguished from each other, but in most cases this requires a rather advanced and specialized photogeological technique. Some of the features which assist in this interpretation are listed below.

1. Clay becomes waterlogged in wet weather and this results in 'puddling' at the gateways to the fields, where cattle, tractors, etc. have passed through. On aerial photographs this appears as pale grey patches adjacent to gates.

2. Clay retains signs of former human activity much better than sand. Consequently ridge and furrow features representing medieval farming are readily seen on clay surfaces but not on sand, and criss-cross plough lines representing several seasons of cultivation are also preserved.

3. Standing water will be common on the clay outcrop; in Cheshire this is generally in the form of small ponds or 'marl holes' dug out by past generations of farmers. Similarly there will be ditches along many field boundaries.

4. Sand contrasts with clay in items (1)–(3) above. In Cheshire hollows (old sand pits) are almost always dry, the topography is frequently hummocky, and the texture of fields underlain by sand differs from that in fields underlain by clay in the absence of plough lines, ridge and furrow, etc.

5. Peat will often show up as dark coloured areas with poor drainage. Networks of drainage ditches constructed by farmers show up clearly on aerial photographs, and the dark peat in the sides of these ditches may be seen. In many cases peat gives poor farming ground and may be covered by woodland dominated by birch.

Survey methods

Artificial exposures

Large artificial exposures such as new road cuttings, sewer and pipeline trenches and sand and gravel pits are the most useful and should be carefully recorded. The two first mentioned will be quickly grassed over or filled in and need to be examined with no delay. It is useful to cultivate contacts with the engineers responsible so that progress reports can be obtained by telephone. Gravel pits constantly yield new information during working and valuable three-dimensional pictures can be built up from regular visits. It is useful to arrange with the manager for any object of possible value (bones, shells, unusual erratics) to be placed on one side. The methods of recording these temporary sections have been described in previous chapters but can be summarized as follows:

1. Careful measurements must be made of working faces and their relations to permanent features recorded by accurate survey.

2. A stereophotographic record of the exposures should be accompanied by field sketches. There should also be close-up photographs where necessary, and accurate locations of any specimens collected must be recorded on the field sketches. Special techniques, such as pebble counts, will be necessary on occasions.

Borehole records

Records should be compiled of all available borehole data. These may be held by the UK Institute of Geological Sciences, by local authorities or by engineering contractors.

Field mapping

If a detailed survey is intended the actual field mapping would normally proceed in the manner indicated below and as illustrated on Fig. 83.

1. The photomorphological map will clearly be revised in the field as some of the photogeological observations are confirmed and others modified. Morphology must be plotted on to a field map, and this can most conveniently be done by means of form lines, which should bring out the positions and shapes of mounds, hollows and slopes. The morphology is of utmost importance to the interpretation of the drift deposits, and details of morainic mounds, kettleholes, eskers, drainage channels and other features should be clearly shown on field maps (Fig. 83). In plotting this information it is desirable to be as objective as possible and consequently form lines are preferred to the interpretative symbols which are used by some surveyors (such as a certain type of arrow for a certain type of drainage channel). This type of interpretative symbol, and also vague representations of broad areas that are characterized, say, by mounds and hollows without showing their positions and shapes are more appropriate to a reconnaissance survey than to the more detailed survey described in this chapter.

Fig. 83 (overleaf) Copy of a field map of glacial deposits on the Cheshire Plain, English Midlands. The area is largely unexposed and lithologies were determined by auger (Yates and Moseley 1967; and Figs 85 and 86). It is important to represent both morphology and lithology on maps of this sort. Note that different lithologies are denoted by letters only, whereas on the original field map coloured crayons were also used. If known locations of each lithology are marked in this way, much clearer pictures will emerge. The inset map shows the location of the Cheshire Plain (C) and of Figs 83 to 88 (asterisk). S, Stoke; M, Manchester. The surrounding ground is stippled.

Fig. 84 (overleaf) Copies of field maps of parts of the Cheshire Plain, English Midlands showing some of the problems encountered in the mapping of lowland drift deposits. Map **A** shows a residual patch of till which forms a flat surface with two hollows. One hollow is dry and has gone through the till into underlying sand, whereas the other hollow containing water is entirely within the till. The hummocky sand surrounding the till is characteristic of a depositional topography, and the peat hollow is part of a kettlehole. This map shows the importance of indicating morphology by form lines. Map **B** illustrates one of the difficulties of mapping. In the north-east sand rests on clay, but in the south the same clay rests on what at first sight appears to be the same sand. In fact a close pattern of auger holes revealed a thin clay of variable thickness that thinned out at one point so that a higher sand came to rest on a lower sand of similar lithology. Some hollows in the higher sand go through into the clay (ponds) and some hollows in the clay go through into the sand (dry) (Yates and Moseley 1967).

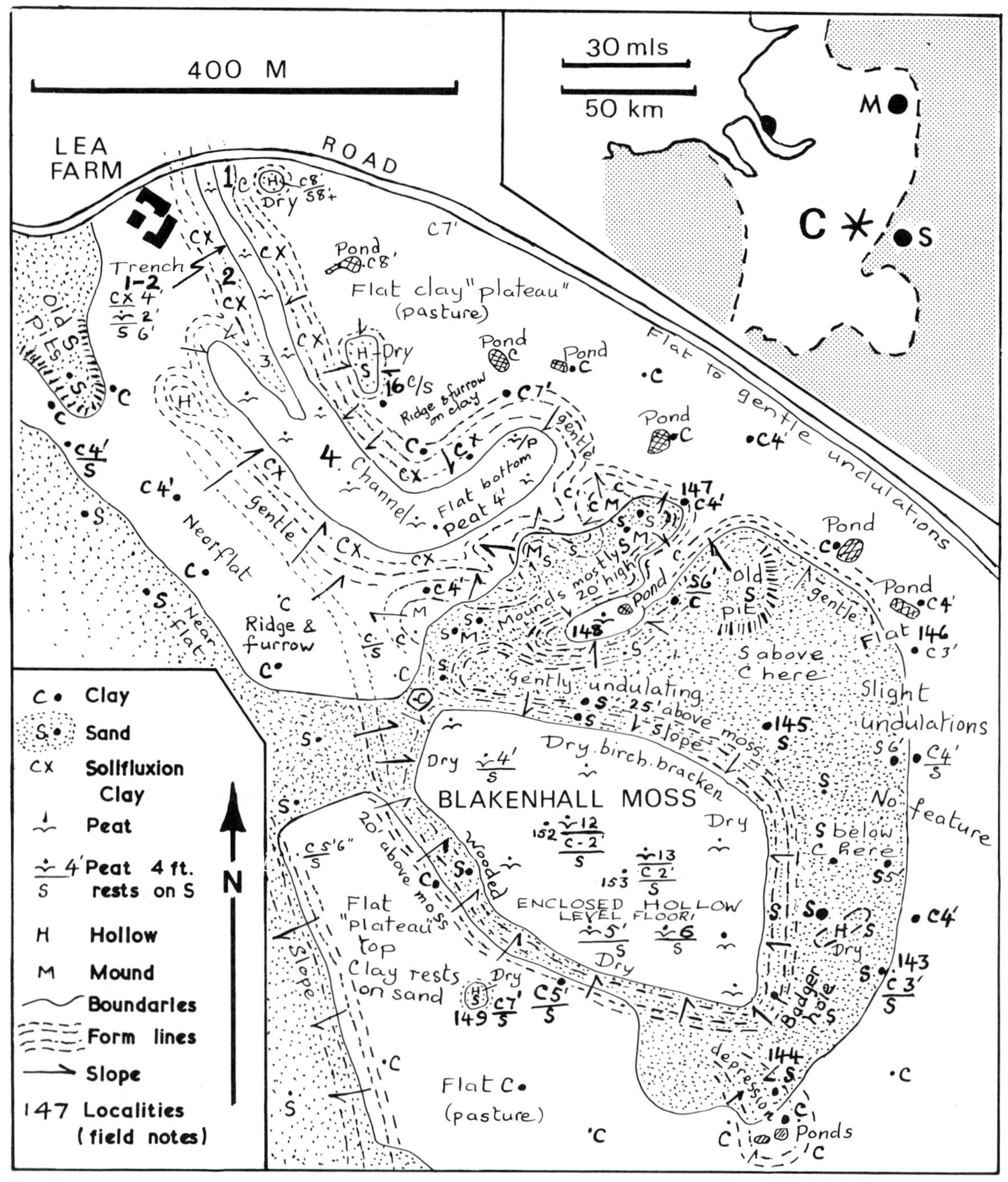
400 M
30 mls
50 km
M
C
S
LEA FARM
ROAD
Trench 1-2
Old S Pits
Flat clay "plateau" (pasture)
Pond
Dry
Ridge & furrow on clay
Flat to gentle undulations
Channel
Flat bottom peat 4'
Gentle
Nearflat
Near flat
Ridge & furrow
Mounds mostly 20' high
Old S Pit
gentle
Flat
S above C here
Slight undulations
Gently undulating
25' above moss slope
Dry birch bracken
BLAKENHALL MOSS
No feature
S below C here
20' above moss
Wooded
Flat "plateau" top
Clay rests on sand
ENCLOSED HOLLOW LEVEL FLOOR!
Badger hole
depression
Slope
Flat C (pasture)
Ponds
Clay
Sand
Solifluxion Clay
Peat
4' Peat 4 ft. rests on S
Hollow
Mound
Boundaries
Form lines
Slope
147 Localities (field notes)
N

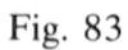

Fig. 83

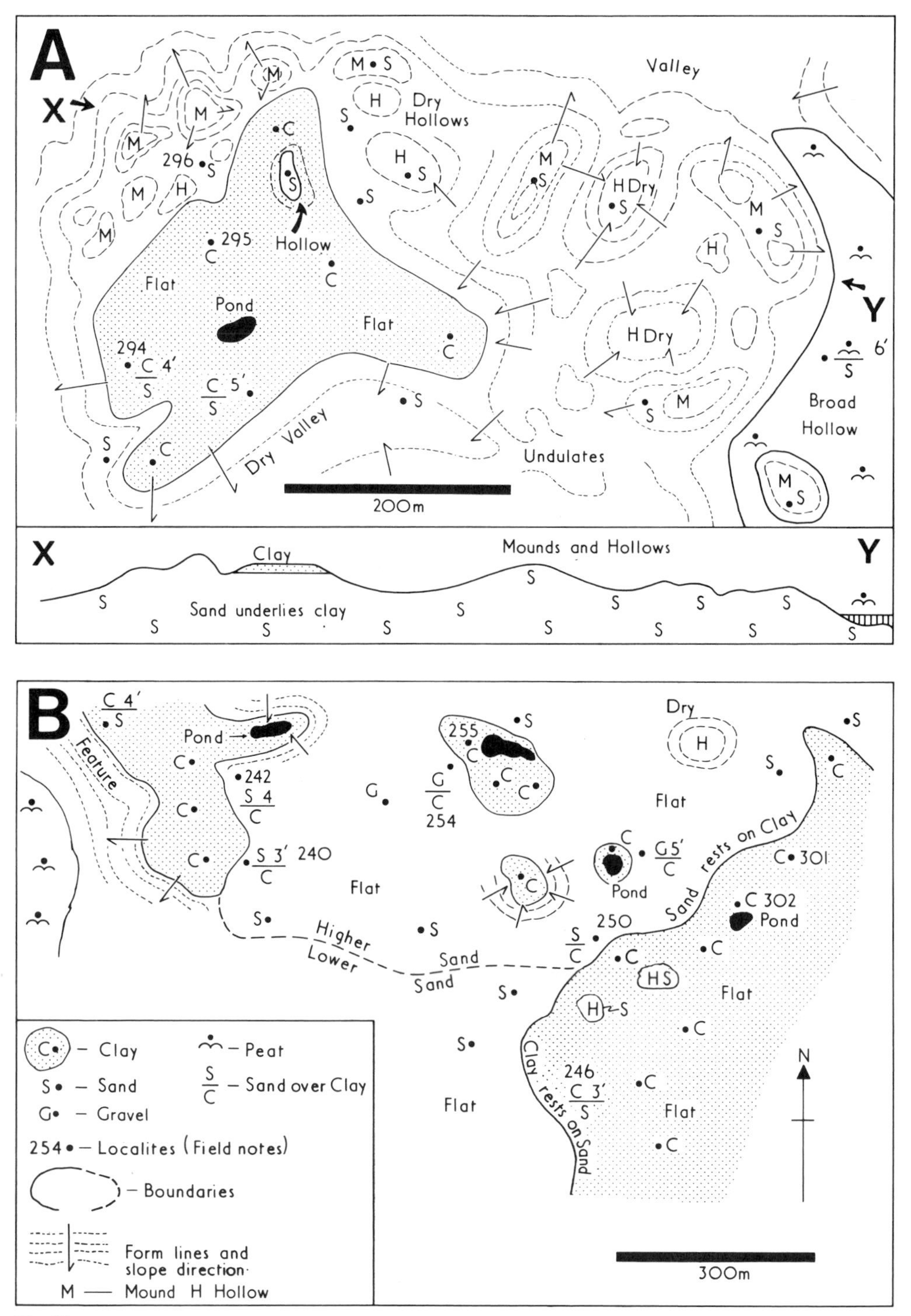
A
X
Y
Valley
Dry Hollows
Hollow
Flat
Pond
Flat
Dry Valley
Undulates
Broad Hollow
200m
Mounds and Hollows
Clay
Sand underlies clay
B
Feature
Pond
Dry
Flat
Flat
Higher
Lower
Sand
Sand
Sand rests on Clay
Clay rests on Sand
Pond
Flat
Flat
300m
N
C – Clay
S – Sand
G – Gravel
Peat
S/C – Sand over Clay
254 – Localites (Field notes)
Boundaries
Form lines and slope direction
M — Mound H Hollow

Fig. 84

2. It is of obvious importance to show the distribution of different types of deposit and most of an area such as the Cheshire Plain will be unexposed, making determinations of lithological boundaries difficult. Obvious use will be made of artificial exposures as indicated above, and of exposures such as badger holes, but these will be limited and most of the ground will have to be examined by using a variety of augers, which fall into several categories, as follows:

A small diameter (3 cm) screw auger with a 1.5 m rod will be part of the everyday field equipment (Fig. 85), just as a hammer is automatically carried when examining solid formations. This enables examination of the surface lithology beneath the top soil at any desired locality. Each of the occurrences of clay, sand, etc. on Fig. 83 represents an auger hole, and in some complicated areas it may be necessary to adopt a close spacing with holes on a 2 to 3-m grid.

Occasionally deeper holes and larger samples will be necessary in which case a 10 or 15-cm diameter auger can be used (Fig. 86). The purity of the required sample will determine whether a bucket auger or a 'sampling' auger is used, the former for lithological determinations only and the latter if the sample is required for more detailed laboratory work such as the separation of a fauna and flora from organic deposits. Extension rods enable the surveyor to put down holes of 5 or 6 m in normal circumstances (exceptionally up to 10 m although assistance would be required for such a project). It is important to realize that these deeper holes need to be planned in

Fig. 85 Survey of the northern part of Fig. 83 by undergraduates (1974). A 4-foot screw auger is used as standard equipment. It is necessary to penetrate the top soil to determine the nature of the deposits beneath. This will determine till sand, peat, etc. on level ground but on slopes these materials may be concealed by solifluction and hillwash clay and sand, which may closely resemble the deposits from which they were derived.

advance since a larger auger plus extension rods and pipe wrenches will be quite heavy, and not the sort of equipment to be carried on a normal field survey. In some cases a motorized auger may be available, saving much energy but it will not generally be accessible to the same range of localities.

3. Using the methods outlined above it should be possible to construct a detailed lithological map, and by concentrating on junctions between one deposit and another, to decide upon the stratigraphical succession. However, the latter is not as easy as it may sound in this sort of terrain; several of the difficulties are listed below.

It is not always easy to draw sharp boundaries between different lithologies which may be gradational or interlayered. For example, the boundary between the gravel and clay on Fig. 84 is difficult to determine because of interbedding, and the outcrops on the surface are complex. Even closely spaced auger holes show apparently random distribution of the two lithologies in this particular case and there seems to be no alternative to a stratigraphical unit composed of clay, sand and gravel alternations as shown in Yates and Moseley (1967, fig. 5). Without occasional artificial exposures interpretations of regions where there are complex alternations of lenses of sand and gravel, typical of morainic areas, can be difficult no matter how closely spaced the auger holes.

Another difficulty illustrated by Fig. 84 arises from what may be called a 'spiral stratigraphy'. For example, auger determinations and other exposures may reveal a clay resting on sand. When this junction is traced across country it may eventually be found that 'the same sand' now rests on 'the same clay' and mysteriously the succession appears to have become inverted. This is clearly a nonsense and in the case just cited means that the clay wedges out, and an 'upper sand' comes to rest on a 'lower sand', with the lithologies so similar that they cannot be distinguished by augering. Interbedding at junctions also makes it difficult to auger from the upper sand through a thin wedging out layer of clay into the lower sand, and this is an occasion when a deeper auger hole should be attempted with the object of proving the threefold sand–clay–sand sequence.

A third difficulty experienced in almost any glaciated area is the redistribution of the unconsolidated glacial deposits in the late and postglacial periods. Solifluction during permafrost times and subsequent hillwash frequently result in the movement of clay and sand down quite gentle slopes. This can be extremely difficult to detect by augering alone, since the deposits differ little from their original form and even good exposures are capable of misinterpretation. Evidence of a recent origin may be found in the over-riding of recent organic deposits and other reliable marker horizons by the soliflucted material,

◁ *Fig. 86* As a survey proceeds some sites will be recorded which require deeper auger holes. The bucket auger with extension rods and pipe wrenches in use here are heavy, and if one is working alone it is hard work to carry them on to site and to put down the hole. A party of Birmingham University students makes light of this however. The locality is on Blakenhall Moss (Fig. 83), where 4 m of peat was found to be underlain by 60 cm of plastic grey clay which was underlain in turn by sand.

and measurements of pebble orientation may reveal a downslope movement inconsistent with a till of glacial origin. On Figs 83 and 87 till-like clay (CX and 1S) rests on postglacial peat and therefore cannot possibly be a till. Where there are good artificial exposures, however, important conclusions are possible concerning the genesis of the deposits whether glacial, fluvio-glacial or postglacial. Fig. 88 indicates one possible application of the sedimentology.

It is concluded that to map successfully a region of glacial deposits such as those of Cheshire it is necessary to record both morphology and the distribution of lithologies, the latter determined largely by augering. Artificial exposures will greatly assist in interpretation and it is a great advantage to have had some experience of modern glacier conditions.

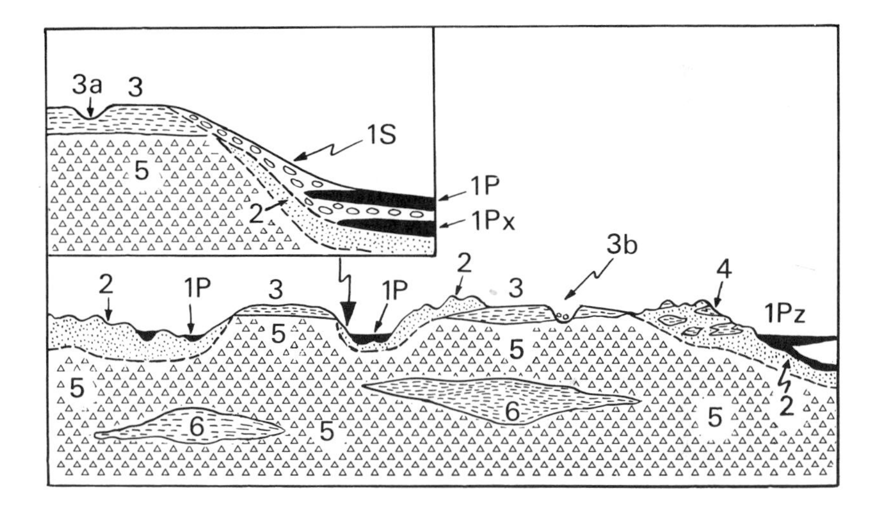

Fig. 87 Diagrammatic sections showing the relationship of different glacial and postglacial deposits in the part of the Cheshire Plain illustrated by Figs 83 to 88. All the deposits represented belong to the last glacial and postglacial episodes. 6, Lenses of till in sand. 5, Glacial sand, mostly outwash, locally below sand (2) and clay (3). 4, Sand–clay (3). 4, Sand–clay complexes with depositional morphology of mounds and hollows. Lateral equivalent of sand (2) and clay (3), also passes laterally into sand (5), which is reworked (interpreted as morainic deposits). 3, Till usually forming flat 'plateau' areas (melt-out from stagnant ice): in part beneath sand (2), in part the lateral equivalent of (2); rests on sand (5). 2, The youngest glacial sand: includes much reworked sand from (5), exhibiting depositional morphology as kettledrift (mounds and hollows); may rest on clay (3) or be its lateral equivalent; may also rest on sand (5) in which case the boundary becomes difficult to determine (see Fig. 85). 1S, Postglacial solifluction and hillwash. 1P, Recent (Flandrian) peat formed in kettleholes and overlain and underlain by till like hillwashed clay. 1PX, Peat dated as late glacial. 1PZ, Swchwingmoor (deep water with Flandrian peat extending across the water surface, e.g. Wybunbury Moss). Formation of 1PZ mostly a result of salt subsidence from natural migration of salt in the underlying Triassic saliferous beds.

This diagram reveals the problems of dating and correlating deposits of this kind. It would be easy, for example, to interpret 1PX as interglacial peat overlain by till of the last glaciation, yet it has a radiocarbon date of 10 780 BP and pollen analysis has shown it to be late glacial (zone III) (from Yates and Moseley 1967).

14 Alpine structures in Mesozoic–Tertiary sediments of the Pre-Betic Cordilleras, SE Spain

Geological survey analogues of SE Spain are to be found in most regions where there is a combination of a Mediterranean type of climate, a predominantly limestone sequence and a mountainous terrain. Geological survey depends on more than the rocks of a region. Vegetation influences, access to exposures, and temperature can be important, especially if work has to be completed during the hot season. Conditions are similar throughout the Mediterranean (Italy, Greece, parts of Turkey, Cyprus, Israel, Algeria, etc.), and in parts of California, Australia and Chile.

The Pre-Betic is the most northerly structural zone of the complex Alpine fold belt of the Betic Cordilleras, extending 600 km from Denia in the east to Gibraltar in the south-west (Fig. 89). It is a largely authochthonous zone of complex structures in Triassic, Cretaceous and Tertiary sediments, the Triassic being mostly ferruginous, gypsiferous red mudstone, which has risen diapirically through the overlying rocks, while the last two are mostly limestones and marls of great variety. The major events of the Alpine Orogeny culminated in the late Miocene, resulting in intense and diverse structures on all scales. Post-orogenic deposits include extensive fanglomerate, tectonically tilted in places, and a great deal of calcrete. Evidence of recent uplift is provided by high structural mountains through which there are impressive antecedent slot gorges, many of them 300 or 400 m deep. The purpose of this chapter is to discuss the problems of mapping rocks with these lithologies and structures in the hot climate and terrain of southern Spain.

Field practice

Undergraduates from northern Europe and particularly from Britain can learn important new methods and gain new experiences in areas of this kind, both from field excursions and by undertaking individual survey projects for final degrees. Some of these can be listed:

1. The region is only 1000 miles from southern Britain, little further than from northern Scotland to London and, being due south, it is the nearest area which is in a completely different climatic zone—almost semidesert in parts.

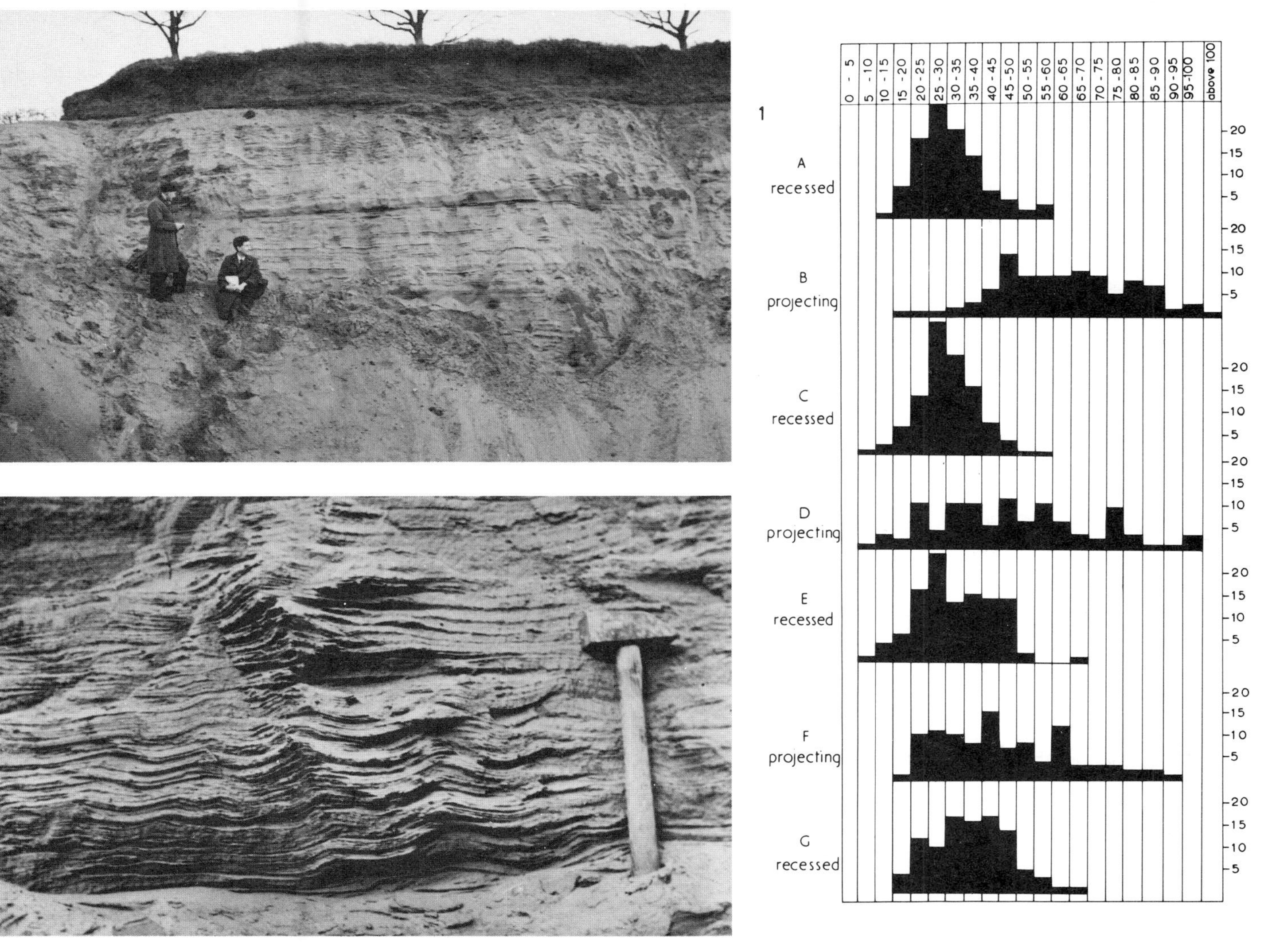

Fig. 88 **A** and **B**, Laminated sands from the Wybunbury Pit about 800 m west of Fig. 83. It is necessary to take care when working in pits such as this, since sand faces can collapse without warning and people have been killed. This is sand 5 of Fig. 87. **C** shows grain size histograms for a sequence of seven of the laminae (four recessed and three projecting). It illustrates one of a number of special studies which can be carried out on these glacial deposits. It will be noticed that the projecting laminae tend to be coarser grained, and may have become slightly cemented because the coarser grain size allowed easier ingress of cementing solutions. Rhythmic deposition is indicated. Grain sizes are in tenths of a millimetre, and frequency percentages are shown (from Yates and Moseley 1967).

2. It is hot in summer and if work is done at this time it can serve as an introduction to geological fieldwork in tropical conditions. Since many geologists will take employment in the tropics it is of some importance that they experience conditions of this kind before plunging into the unknown. It can easily decide on a future career.

3. There are extreme vegetation contrasts with the most northerly latitudes. The mountainsides at first sight seem quite open with easy access, but this usually proves to be an illusion. There is often a deceptive cover of *garique* (*tomillaros*), a waist-high thicket of holm oak, gorse, juniper, fan palm and other shrubs, almost all of which are thorny. Old cultivated terraces, seen from the distance, are most misleading. The shrubs mentioned above have taken over from the vines and vegetables of former years. The terraces were constructed on most mountainsides in gullies almost to the mountain summits, and generally have dry stone walls 2 to 3 m high separating one terrace from the next. The paths which used to connect them (or stone ladders—stones sticking out of the walls as steps) have nearly all decayed, and to climb a 3-m high crumbling terrace wall into a mass of gorse followed by another similar terrace wall 10 m later at the next terrace

Fig. 89 Generalized structural map of the Alpine fold belt of SE Spain. The principle tectonic trends are shown by lines and the Triassic diapirs by fine stipple. The locations of the figures are also indicated.

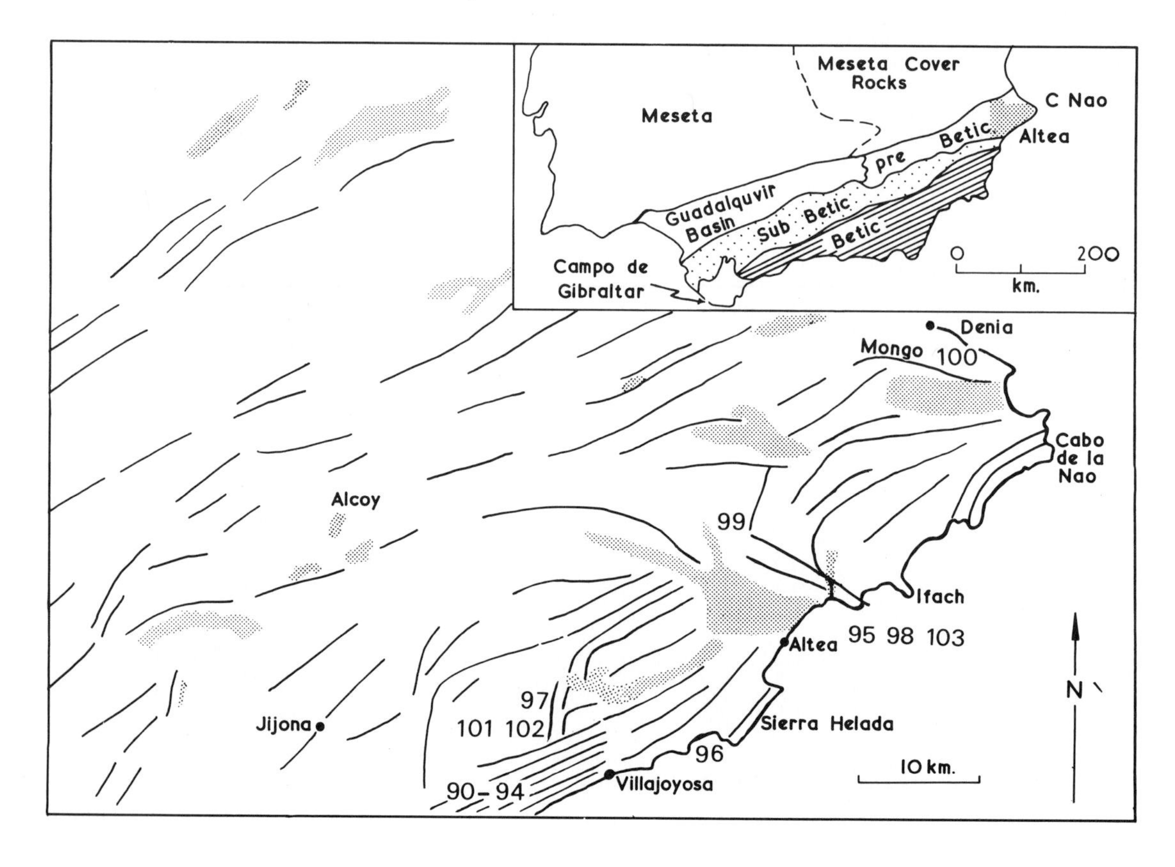

level (and so on for the next 100 m of hillside) quickly teaches the would-be surveyor that a direct route from A to B, up what may appear to be the most convenient slope, is not necessarily the route to follow. It is possible to spend most of a day enmeshed in this vegetation and to make scarcely any geological observations. The answer is experience and careful planning of traverse routes. There are well-concealed paths in most areas which lead through the thicket in minutes compared with the hours it would take to bulldoze through it. Experience of field conditions of this kind can only be helpful in the future should one be faced with (for example) the thickets of East Africa or tropical forests.

4. In Britain one is fortunate to have 1:10 000 or 6 inches to the mile topographical maps with a high standard of accuracy. These serve as valuable base maps. In addition the whole of the United Kingdom is covered by aerial photographs to scales of 1:12 000 and larger. For our work in Spain the largest scale map was 1:25 000 (and these were no more than enlargements of 1:50 000 maps) and the only aerial photographs we could obtain were on a scale of 1:40 000. The latter are of excellent quality, but it is clear that with smaller-scale base maps and photographs, different mapping techniques have to be used compared with traditional British methods.

5. Although in northern Europe most undergraduates can examine wide varieties of rock and structure within a few hundred miles of their home university, they are unlikely to find anything remotely like those to be seen in SE Spain. This particularly includes the Quaternary deposits, recent tectonic activity, weathering and erosion.

It is therefore clear that to do geological work in this region it is not enough merely to be fortified by methods evolved in the London Basin or North Wales. A new technique has to be acquired. All the work I shall refer to in this chapter was done in the summer months, but spring and winter are perfectly suitable. In January it is generally sunny, with the temperature about 15°C.

The climatic conditions during the fieldwork described below can fairly be described as 'tropical'. The work was conducted from tourist camp sites on the coast and, until 'acclimatization' was achieved, generally started early (5.30 a.m.) and continued until about 1.30 p.m. As one becomes used to the heat at midday, it is possible to continue later. It is also most important not to become dehydrated, and to drink as much as one can before starting. If the day's work involves climbing up and down steep slopes a 1-gallon water bottle is necessary. This is heavy but does at least maintain a steady rucksack weight; as the water is consumed, specimens are collected. Individuals will, of course, vary in their need for water. As for dress, in spite of the prickly vegetation I find it most comfortable to work in shorts and a wide-brimmed hat.

Sun glasses and sun cream, especially until one has acclimatized, may be needed. Boots are generally essential and I recommend the oldest, no longer waterproof pair, not a cheap variety of 'desert boots', which will be torn to shreds within a few days on the serrated limestone blocks—weathered to knife-like edges under the prevailing type of weathering. A new pair of quality boots is unlikely to be waterproof on return from a season's summer fieldwork in southern Spain's dry hot conditions. The remaining field kit will be standard, as described in chapter 2.

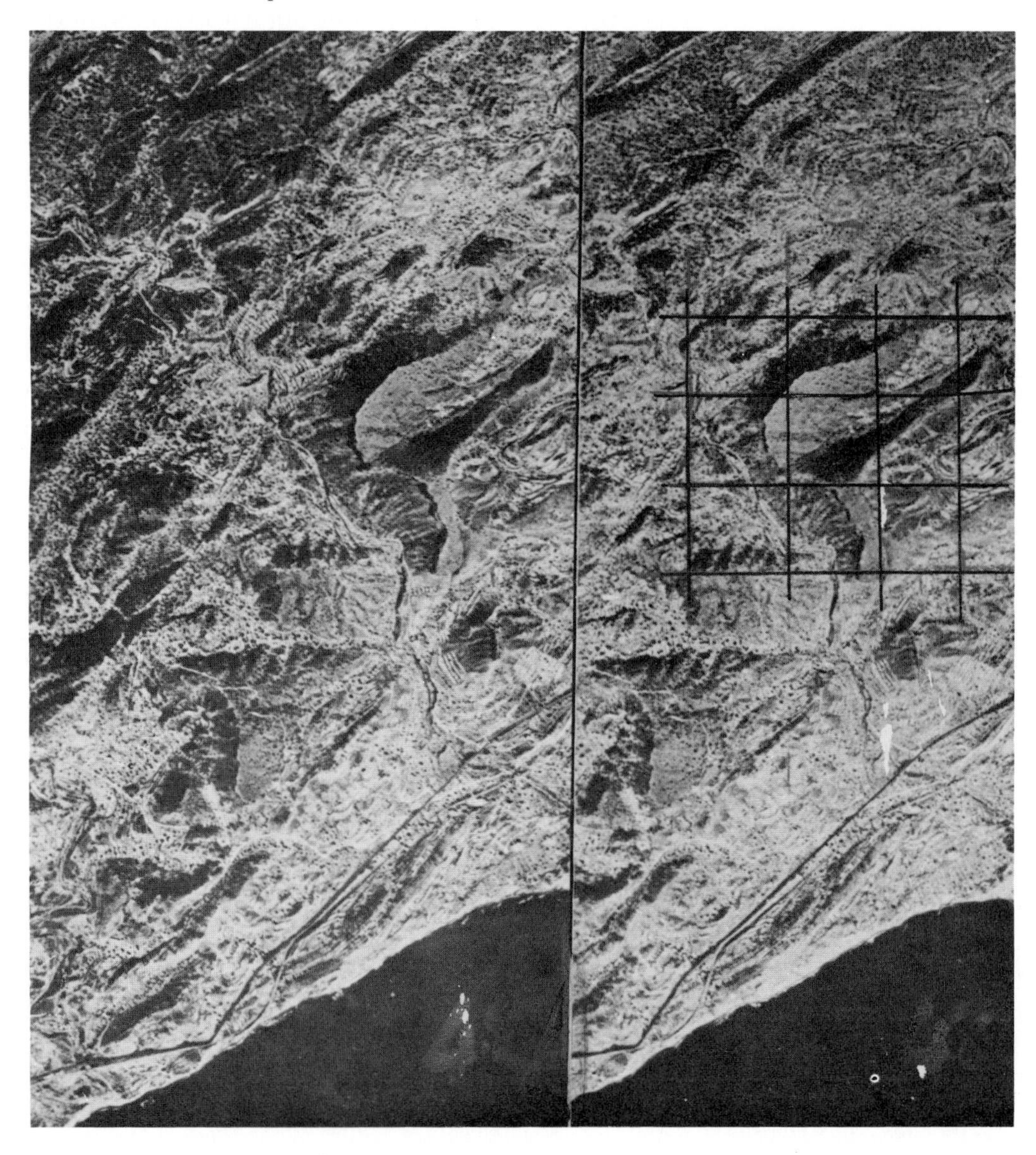

Fig. 90 Stereopair of part of the Villajoyosa fold belt, SE Spain. The method of enlarging part of the photograph so that field information can be plotted is shown. Squares are drawn on the photograph and enlarged as in Fig. 91. (Comision Nacional de Geologia, España.)

Survey methods

The terrain and the small-scale maps and aerial photographs present problems when a detailed geological survey and a large-scale map are intended. It is impossible, for example, to use the traditional British method of plotting geological boundaries and observations on to 6 inch or 1:10 000 maps as indicated in chapter 11 (Carboniferous of northern England). The methods suggested here, however, are appropriate to the greater part of the world, for example much of Australia, Africa and India. The aerial photographs (always viewed stereoscopically) provide the most useful 'basemaps', with principle localities (villages, road intersections, etc.) checked against the 1:50 000 and 1:25 000 maps. Unfortunately the aerial photograph scale at 1:40 000 is much too small and it is necessary to have enlargements. These can be made in two ways:

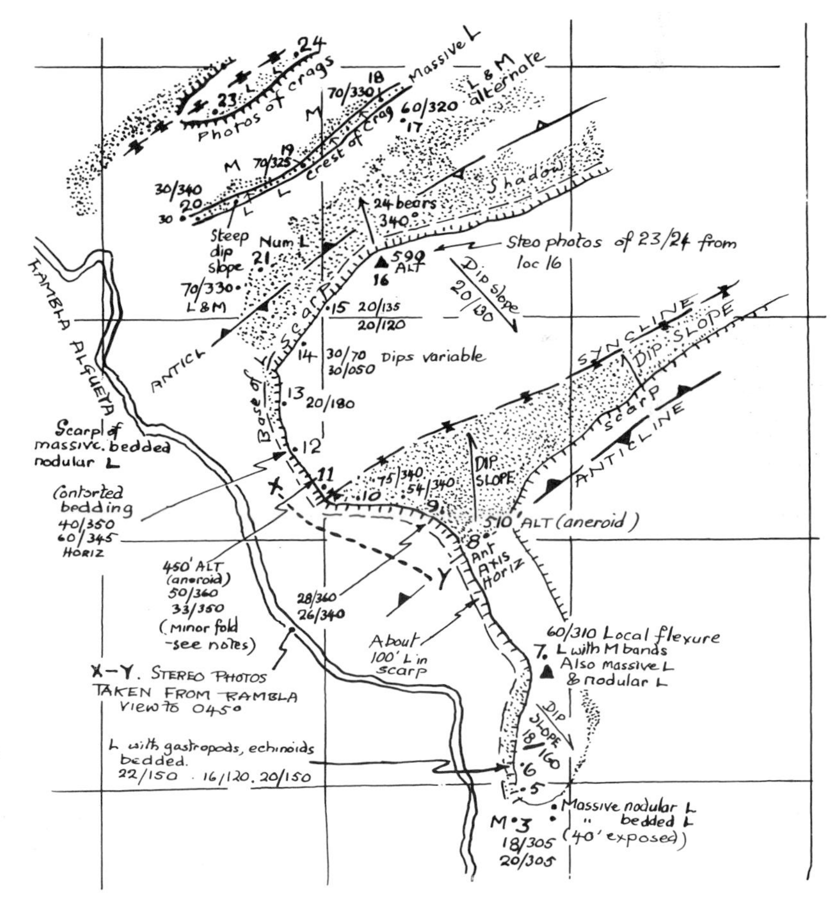

Fig. 91 Field map of part of the Villajoyosa fold belt, SE Spain. The principal features seen on the aerial photographs were enlarged from squares on the photographs (Fig. 90). They include the watercourse (rambla), the escarpments and areas in shadow. The latter shown as stipple was represented as light crayon shading on the original. The enlarged base map made it easy to plot geological information in the correct positions. Although the basic structure, shown by the section on Fig. 93, is simple, this field map shows complex variations in dip (see Fig. 92).

1. Photographs of areas already known to be of great interest, perhaps from work of a previous field season, can be enlarged to, say, 1:10 000. This is possible with the Spanish photographs, which are of good quality, but inevitably there is loss of detail. However, when these photographs (or photocopies of them) are used as maps field observations can be plotted on to them at the larger scale. The original photographs provide the check for photogeological detail, which can be transferred to the enlargements. It is important to remember that stereoscopic enlargement is always better than photographic enlargement of negatives from the original photographs, but the former is not good for plotting details since the space for writing is too small.

2. The second method is the more useful, especially for undergraduate projects when areas of exceptional interest will not usually be known beforehand (Figs 90, 91 and 93). A 1-cm grid is drawn lightly on the aerial photograph with an all-purpose (wax) pencil. An enlarged grid, to whatever scale required, is then drawn on a mapping sheet (field notes) and obvious topographical features (water courses, roads and tracks, escarpments, prominent summits, buildings, large trees etc.) are transferred from the photograph to this map. There is then plenty of space for plotting geological boundaries, structural and petrological observations, locality numbers and so on. These field maps can later be combined to provide the final map.

These methods overcome the problems of small-scale maps and photographs and detailed geological maps of most areas can easily be produced. To reiterate, the precise positions of geological features and locations are located stereoscopically on the small-scale photographs and these are transferred to the larger-scale plan.

Geological problems

There may also be geological problems in a new area. The geology of southern Spain, for example, differs in most respects from that of Britain and northern Europe.

Palaeontology

Before starting on a survey of this kind, it is of course important to be acquainted with the fauna and flora likely to be encountered. During a first season of mapping, and undergraduate projects are only one season, identification to generic level may be possible in the field, but specific names will be out of the question; not even generic names will be possible for the many foraminifera that can only be seen with a hand lens. The important point is to collect numerous accurately located specimens, and not to worry too much about field identification. In some cases where the fauna is abundant, sufficient specimens can be

Fig. 92 Diagrams drawn from field sketches and stereophotographs from locality 16(A) and position B(X–Y) on Fig. 91. These diagrams illustrate unusual structures in that there is a massively bedded nodular limestone (the Charco Limestone) which is folded into relatively simple larger structures as seen on Fig. 93, and indicated by the base of the limestone on 4X–Y, but in detail is contorted into numerous complex minor folds. The latter are almost certainly caused by penecontemporaneous gravity slumping, but are misleadingly like tectonic structures when first encountered.

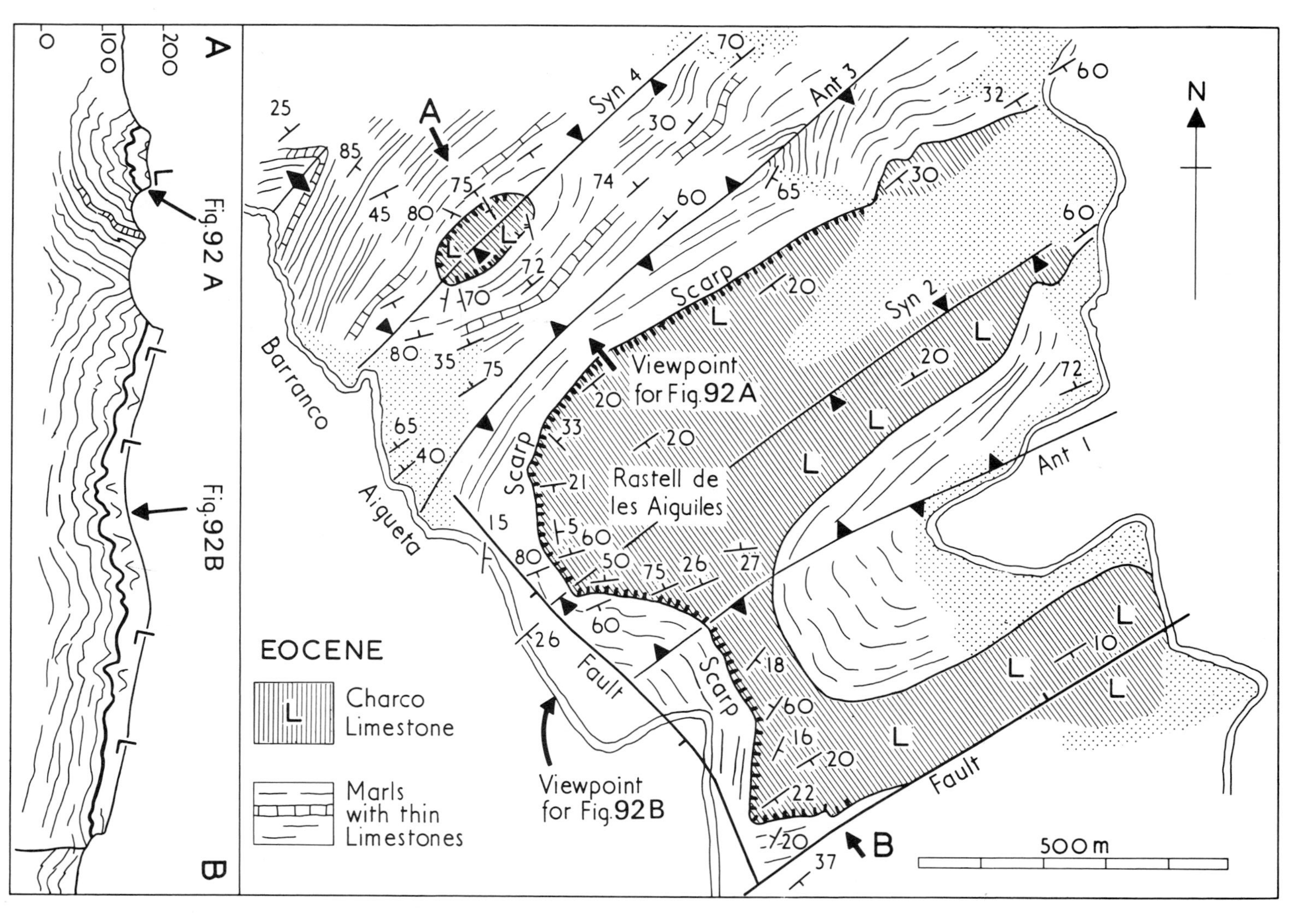

Fig. 93 Final map of the area of Fig. 91, drawn up for publication from field notes and maps. The series from Figs 90 to 93 indicates the procedures to follow during an investigation of this kind where no large-scale topographic maps are available.

collected, horizon by horizon, for statistical studies; for example, D. Stevens collected several hundred echinoids when mapping the Amadorio Dome near Villajoyosa during an undergraduate project. More precise identification can be done on return from the field; for example, examination of thin sections and polished sections, consultation of monographs and with university staff who specialize in the particular fossil groups, and perhaps even visits to museums where there are type specimens of the species collected.

Sedimentology

Most rocks of southern Spain are sedimentary, and the majority of these are carbonates. Like other branches of geology, sedimentology has become a specialized subject. The terminology is continuously being modified, and undergraduates at the end of their second year are unlikely to be familiar with all the terms and concepts now in use. Before going into the field they should have some working knowledge of modern terminology so that the field observations and collections can provide sufficient information for more accurate determinations later. In southern Spain numerous field terms, such as shell limestone, nodular limestone, cementstone, marl and pebbly mudstone, can be used but alternative names may be preferred when the project is written up. It is necessary to take close-up photographs of all lithologies and to collect specimens, of which a few will be large, but the majority—several hundred—will be small. The small specimens can later be laid out in the laboratory and examined using a binocular microscope; some characteristic ones can be sectioned and final determinations of rock names can then be made. There will be other sedimentary structures, some familiar and some not, which will normally be too large to collect and these should be sketched and photographed (Figs 92, 94 and 95).

Calcrete deserves a special mention. This is a common sedimentary rock in this part of the world (Figs 96 and 139) but is seen in Britain only in the desert formations of the Devonian and the Permo-Trias. It is found in all the semiarid subtropical regions of the world—all the Mediterranean lands, large parts of Australia, India, East Africa, Brazil, Texas, etc. It is mostly a Quaternary deposit formed by precipitation of calcium carbonate in the subsoil, which in high rainfall regions is removed completely in solution. Eventually a hard limestone 'pan' forms in the subsoil. This is often underlain by a pisolitic or friable zone that is less well cemented. Erosion may strip off the thin topsoil leaving an extensive limestone pavement, not unlike the 'clint and gryke' pavements of the Yorkshire Carboniferous Limestone (Fig. 47). Screes can be cemented in the same way, and it is common for geologists who have never before seen these deposits to be completely mislead and think of them as older 'solid' formations.

Fig. 94 (overleaf) Pebbly mudstones of Eocene age, one of the common rock types in SE Spain. These lithologies are best recorded by photography, field sketches and the collection of specimens. Laboratory examination of the latter revealed large numbers of foraminifera. The rock is completely unsorted with angular pebbles of limestone and deformed fragments of pale siltstone (clearly unconsolidated at the time of deposition) set in an argillaceous matrix. These deposits are believed to have been formed by submarine debris flow down gentle slopes, perhaps triggered by earthquakes.

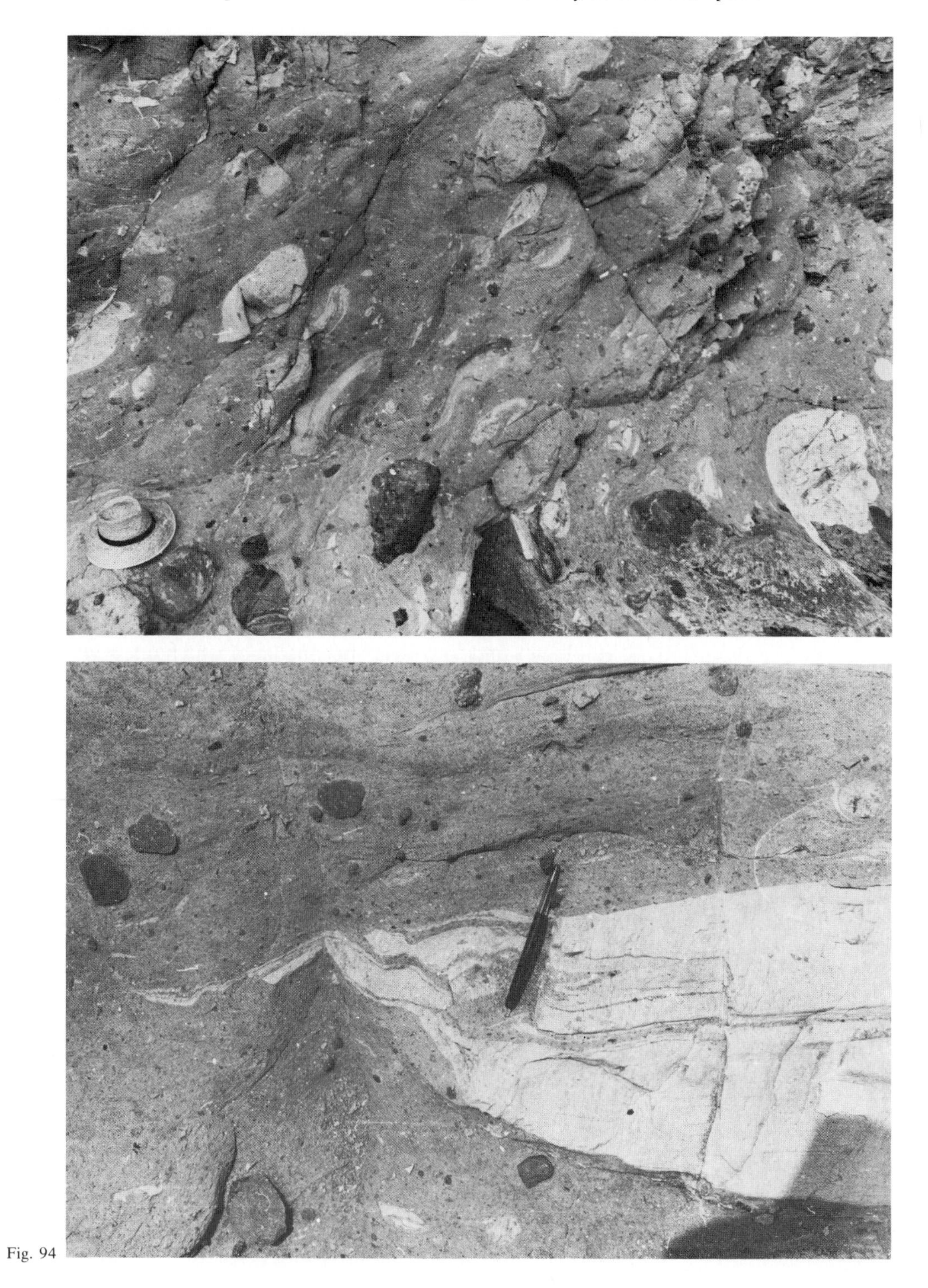

Fig. 94

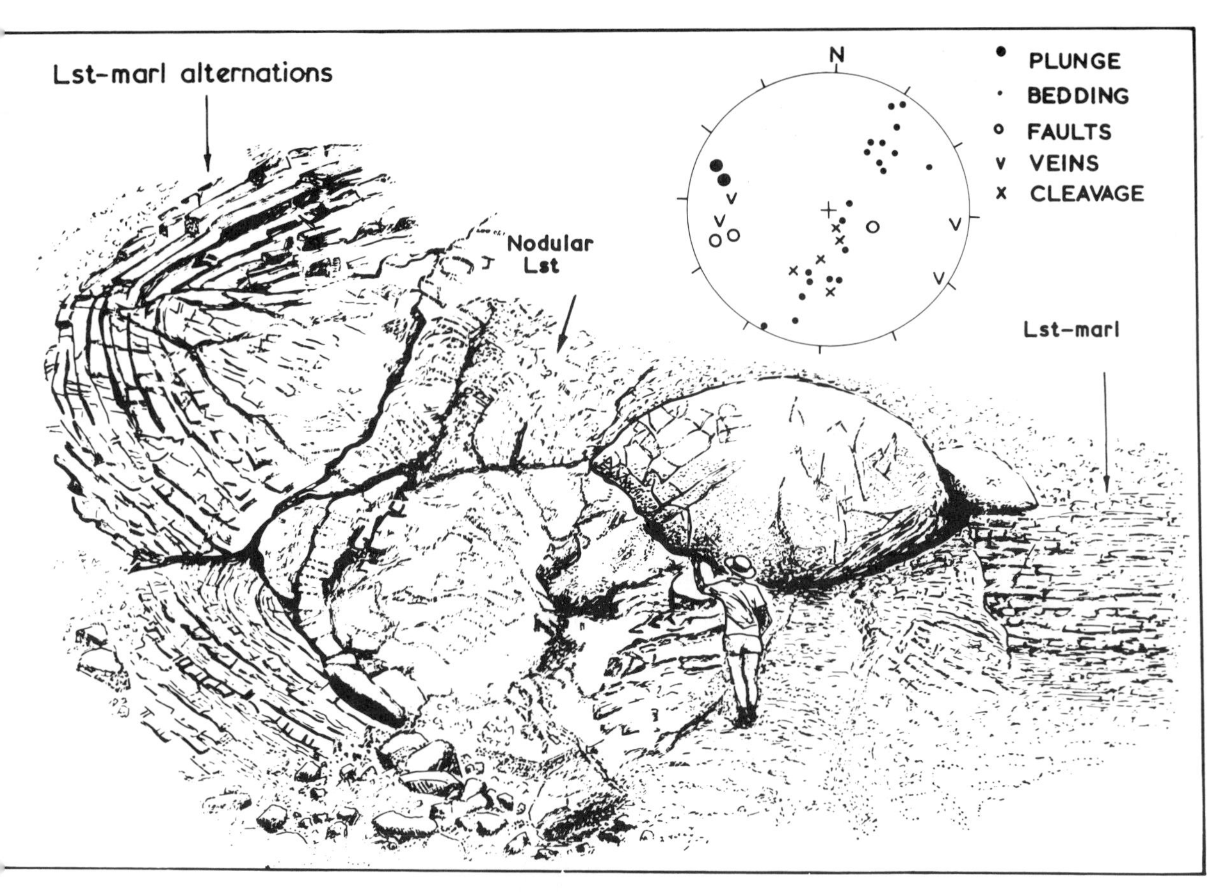

△
Fig. 95 A section on the Mascarat shore (SE Spain) in Eocene rocks showing: (1) a recumbent fold in marl–limestone interbeds, and more massive nodular limestone for which the structural data have been plotted stereographically; and (2) a large septarian nodule. Diagrams of this type can show structures pictorially, with their detailed three-dimensional orientations plotted on stereographic projections (Moseley 1973).

Fig. 96 Recent calcrete, Benidorm, SE Spain. There is a hard massive limestone cap rock, underlain by a rubbly friable zone. Calcrete (caliche or kunkar) frequently forms limestone pavements, and can easily be mistaken for an older limestone formation (Moseley 1965). Compare with Figs 47 and 139. ▷

Structure

The structures of the Pre-Betic Zone, although entirely uninfluenced by metamorphism, are exceedingly complex. The Mesozoic and Tertiary sedimentary rocks have been folded along ENE axes, with local abrupt trend changes, mostly to N–S and ESE. The underlying Triassic rocks, predominantly ferruginous, gypsiferous mudstones, have risen diapirically under the influence of the Alpine stresses, the trends of the diapirs being controlled by wrench faults in an underlying basement. These diapirs push aside the younger sediments, adding another structural episode to the already polyphase structures. Both the NS and the ESE folding are a consequence of these events and gravity slides are among the structures to be seen (Moseley 1973). The climax of the orogeny was late in the Miocene, but there has been a sequence of minor Earth movements since the early Tertiary and minor tectonic activity has continued until the present day. These later movements are shown in a dramatic way by the impressive antecedent gorges, which are up to 400 m deep, narrowing to dark 2-m wide, water-filled clefts at the bottom (Figs 97 to 99), and by the 30 to 40° dips on some of the Quaternary gravels seen near the Triassic diapir margins (Warbrick, unpublished Birmingham University report). Interpretation of all these structures is not easy but it can proceed in several stages as follows:

Fig. 97 Part of the antecedent Amadorio slot gorge, a 300-m deep cleft, nearly 2 km long, across a major anticline in massive Cretaceous limestone. The gorge is difficult to traverse, swimming and climbing being necessary in turn, and it illustrates one of the difficulties of fieldwork in SE Spain (Birmingham University field party).

1. The aerial photographs show the broad outline. Most of the major and many of the minor folds can be picked out, some of these being illustrated on Fig. 90. (The various methods of photogeology are described in chapters 3 and 4.)

2. Views of distant mountainsides or opposite valleysides frequently reveal important structures in cross-section. These should be sketched and photographed for stereo-viewing. Again these methods have been described in Part I (Fig. 30; see also Fig. 92).

3. The localities which appear to be important from (1) and (2) above should be visited. Measurements of bedding dips, axial planes and other structural features should then be recorded under locality numbers as indicated in chapter 7 (Fig. 31). In order to determine the true nature of the fold structures numerous dip readings should be taken. Many of the folds, for example, may prove to be conical (Moseley 1968b; see also Hamon 1961 and Stauffer 1964), but in order to demonstrate this a minimum of 20 or 30 dip readings on one small-scale fold (Fig. 100) are necessary.

I have not mentioned all the structural problems likely to be encountered—there are particularly awkward ones associated with the Triassic diapirs for example—but this book is not an appropriate place for a discussion of these.

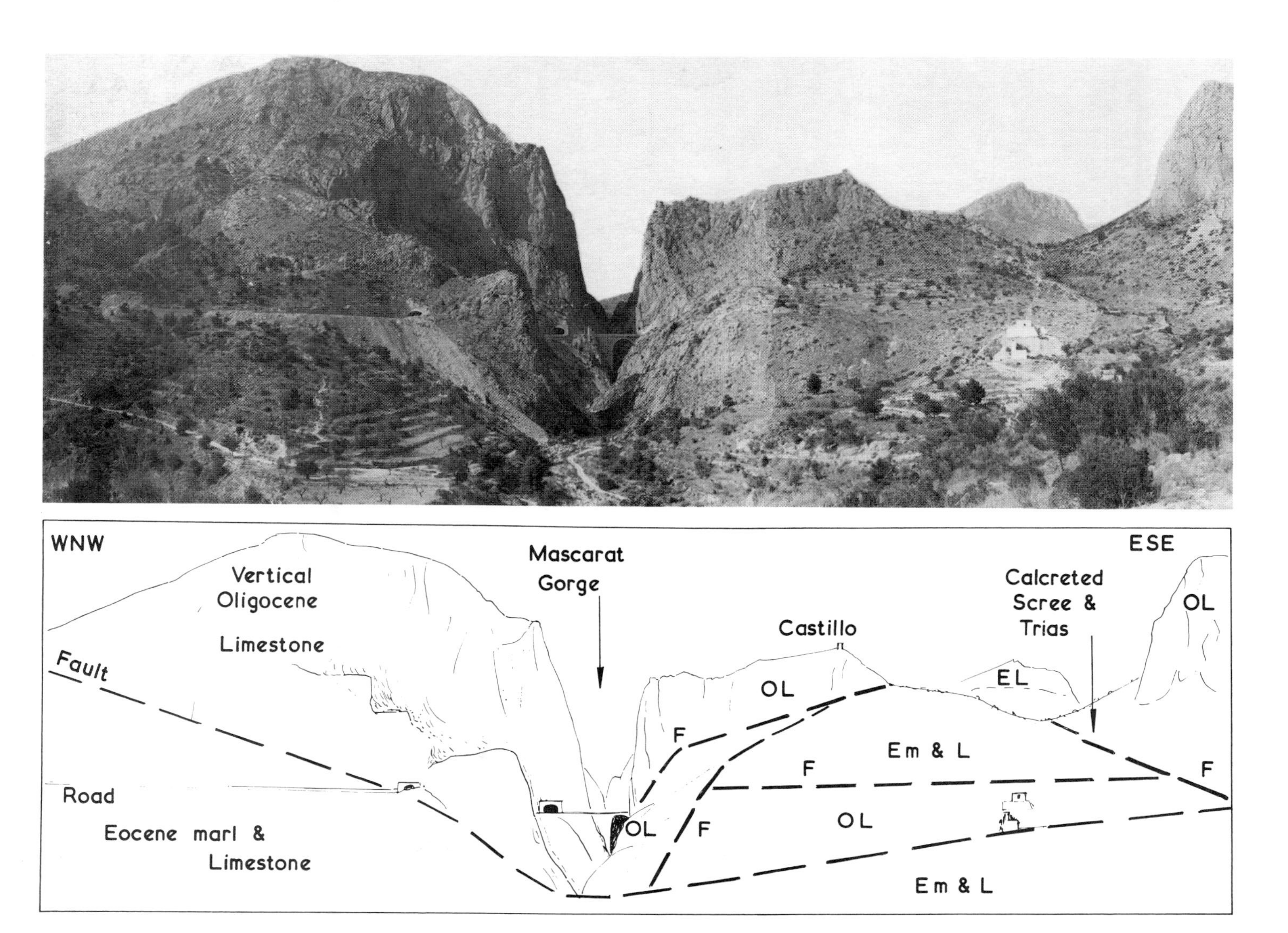

Fig. 98 The Mascarat Gorge (SE Spain), an antecedent cleft across the Bernia uplift, which forms a high ridge extending to the sea. This structure has proved an inconvenience to communications, and the railway, main road and autopista pass through it in tunnels. EL, Eocene limestone; OL, Oligocene limestone; EM, Eocene marl; F, fault.

Fig. 99 A student field party at the Paso de los Bandoleros (SE Spain), a 400-m andecedent slot cut across vertical Cretaceous and Tertiary strata. C, Cretaceous limestone; O, Oligocene limestone; M, Miocene marl with limestone; F, fault.

Fig. 100 Structural lines (strike and dip) for the eastern part of the Mongo Syncline, SE Spain. The summit of Mongo (751 m) and the road junction San Antonio–Denia–Javea are shown on the main map. San Antonio (A), Cape Nao (N) and Sierra Bernia (B) are shown on the inset stereogram. X shows contoured bedding poles, recalculated from an equal area to a Wulff net, with the best-fitting small circle plotted. Contours are at 8 and 4%. Stereogram Z is of bedding poles from two en-echelon folds in the axial region of the syncline, with the best-fitting small circle plotted (Wulff net). This figure shows that a considerable number of bedding plane orientations have to be measured if the true three-dimensional nature of a fold is to be determined. In this case the small circles reveal the presence of conical folding (Moseley 1968b). ▷▷

Stratigraphy

The work on the palaeontology, sedimentology and the structure of the region will help elucidate the stratigraphy which is a combination of all these studies. In the field local rock sequences can be measured by a variety of methods, most of which are illustrated in Part I. For a tape and compass survey a 30-m tape is obviously useful, but it is possible to improvise. I have on occasions marked out 10 feet on a cane picked up as driftwood on the beach and used this to measure both horizontal and vertical distances, and on another occasion measured a

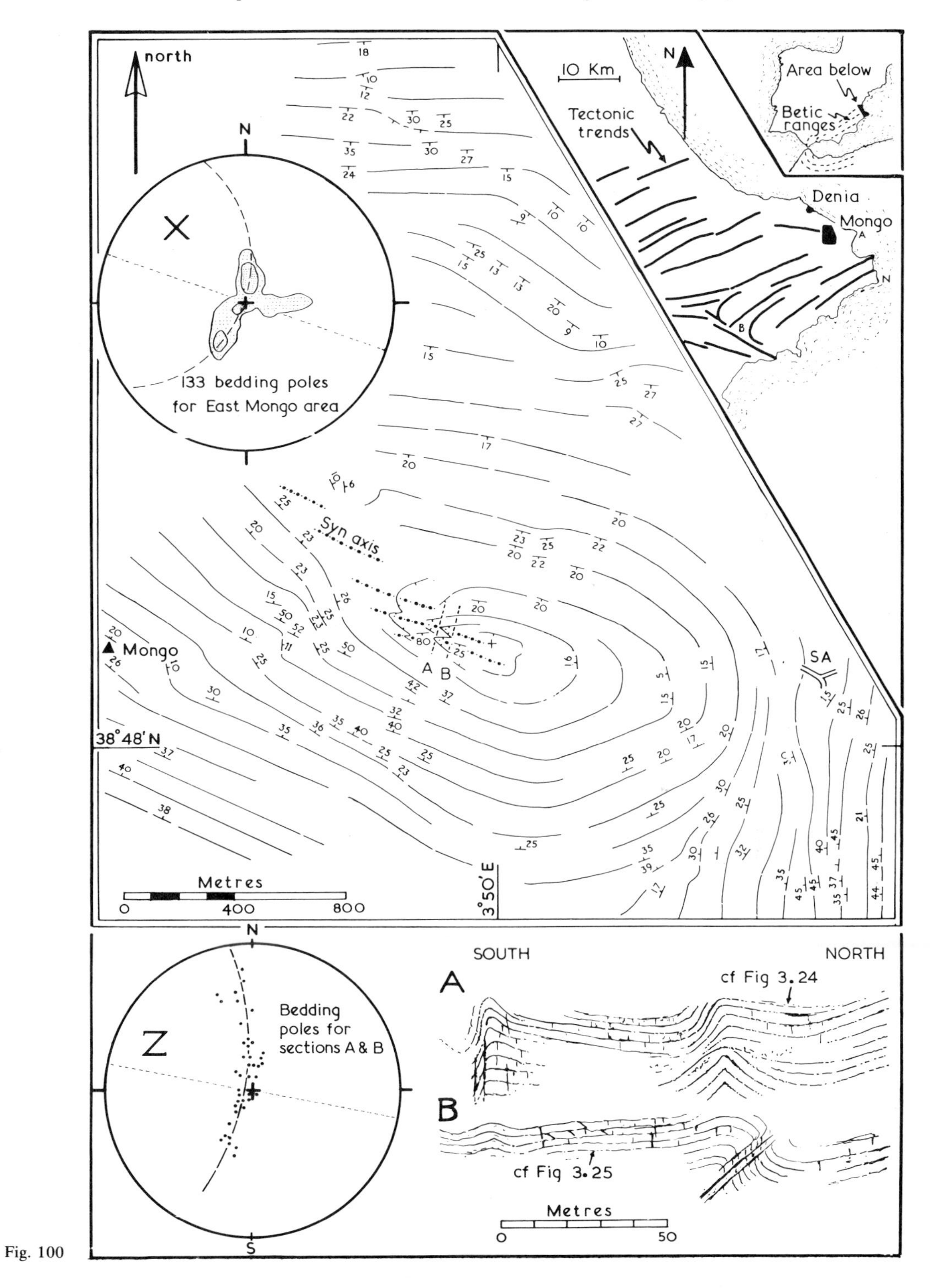
north
N
133 bedding poles for East Mongo area
10 Km
N
Tectonic trends
Area below
Betic ranges
Denia
Mongo
Syn axis
Mongo
SA
38°48'N
3°50'E
Metres
0
400
800
N
Z
Bedding poles for sections A & B
S
SOUTH
NORTH
A
cf Fig 3.24
B
cf Fig 3.25
Metres
0
50

Fig. 100

cliff height using a ball of nylon string with a stone on the end. It is also possible to estimate the height of a vertical cliff by sighting a 45° angle to the top and measuring the horizontal distance to the base. The detail and accuracy of the measurements will, of course, depend on the accuracy required for the succession.

In this region there are several important problems to be overcome some of which are unlikely to be solved during an undergraduate survey.

1. In a highly tectonized region true stratigraphical boundaries between one major unit and another are uncommon; for example, in the Amadorio Dome (Figs 101 and 102, and Stevens, unpublished Birmingham University report) Albian marls with limestones are faulted against massive limestone with occasional fossils of possible Cenomanian age. The problem is how to erect a stratigraphical succession when an indeterminate part of the sequence has been faulted out. This is referred to in the legend to Fig. 101.
2. There are continuous stratigraphical sequences from the Oligocene to the Miocene, although local small angular unconformities can be seen, certainly within the Miocene. However, in the Bernia region (Fig. 103) and elsewhere, the junction is taken at a strong facies change from massive limestone to glauconitic sandy marl and argillaceous limestone (Moseley 1973). This sequence is fossiliferous throughout, but until detailed faunal studies are completed across the boundary, it will not be known whether it represents the true internationally recognized system boundary, or is merely a convenient facies change.

In conclusion it will be evident that this is a region requiring knowledge of palaeontology, sedimentology, structure and stratigraphy; it has important and misleading Quaternary desposits as well as structures of this age. Like practically all geological surveys it is multidisciplinary, and the extreme specialist in any one field cannot hope to solve the problems.

Fig. 101 Map of the Amadorio Dome region near Orcheta, SE Spain. There are two fold trends in this region, ENE and N–S, the latter being caused by diapiric uplift of nearby gypsiferous Trias. Here and elsewhere in SE Spain complete stratigraphical successions can be difficult to determine because so many boundaries are tectonic. It is necessary to determine the sequence for each fault block, and then, hoping to find easily identified marker horizons, to match up one block with another. In this case there are three areas, north, centre and south, separated from each other by faults but with no lithological or faunal horizons in common. It is thus necessary to date the rocks from their contained fossils. In the south *Nummulites* and other fossils clearly show the rocks to be of Eocene age. In the central area the marls (Cm) contain an abundant echinoid fauna of Albian age. The area to the north (C and Cz), however, has so far yielded few macrofossils, although there are foraminifera which suggest (although with no certainty) a Cenomanian age. It is therefore possible to build a tentative sequence of Cm–Cx–Cy–unknown (faulted) gap–C–Cz–unknown (faulted) gap–e. This region has recently been mapped by D. Stevens (Birmingham University) as part of a BSc thesis. His map contains much more detail than is shown here. ▷

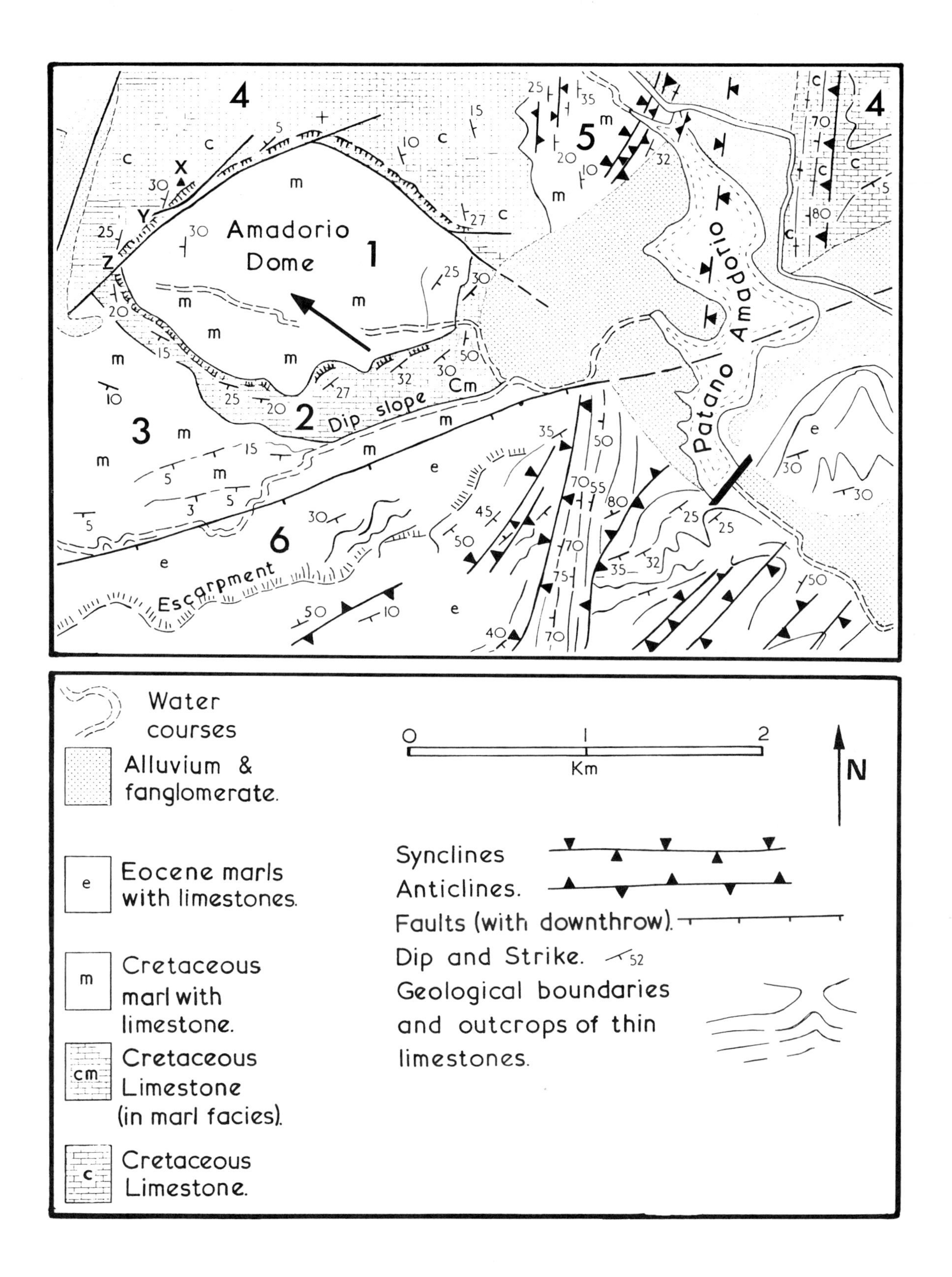
Amadorio Dome
Dip slope
Escarpment
Patano Amadorio
Water courses
Alluvium & fanglomerate.
Eocene marls with limestones.
Cretaceous marl with limestone.
Cretaceous Limestone (in marl facies).
Cretaceous Limestone.
Km
N
Synclines
Anticlines.
Faults (with downthrow).
Dip and Strike.
Geological boundaries and outcrops of thin limestones.

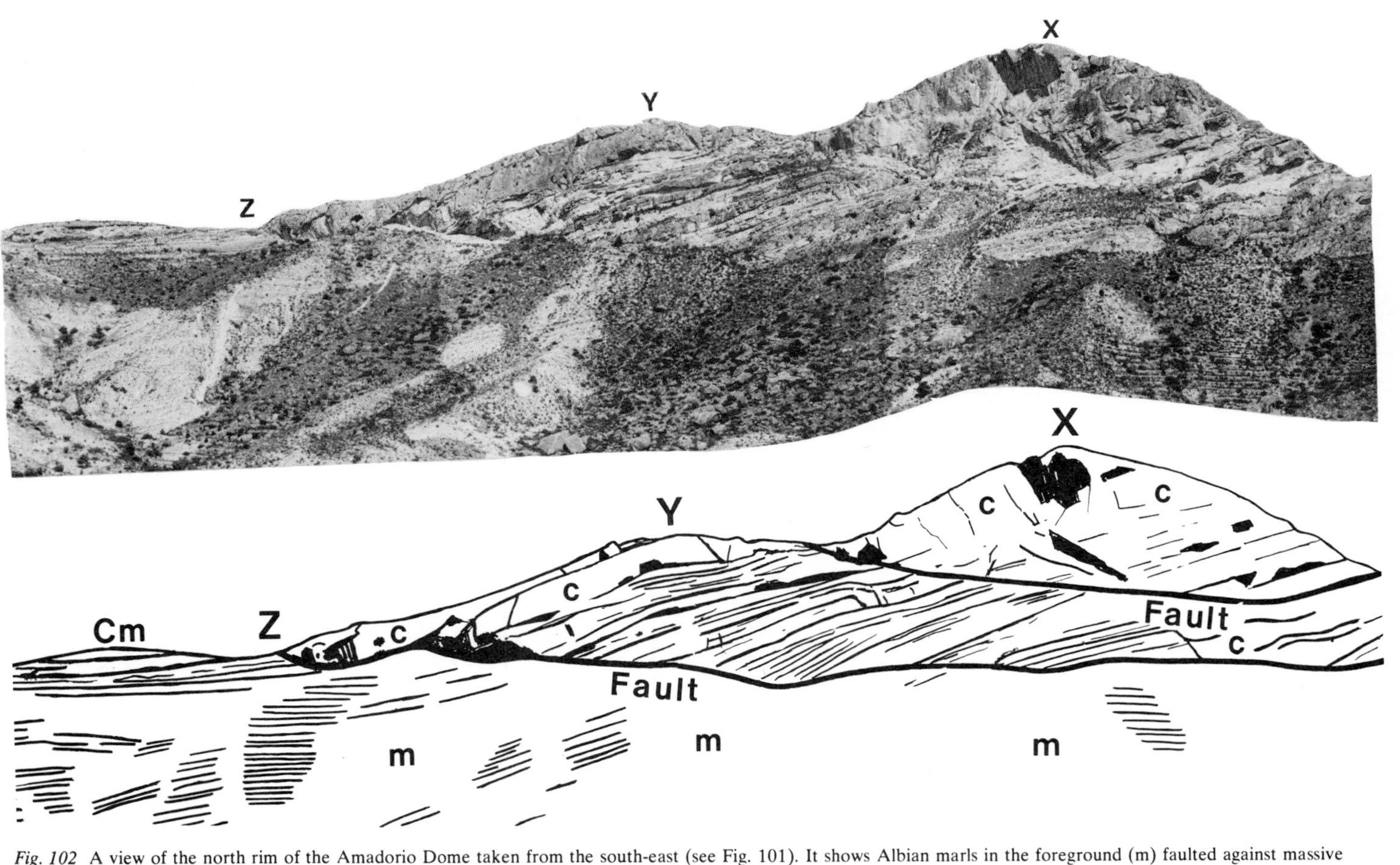

Fig. 102 A view of the north rim of the Amadorio Dome taken from the south-east (see Fig. 101). It shows Albian marls in the foreground (m) faulted against massive and bedded limestone of probable Cenomanian age (c). Many photographs of this type should accompany a survey of such a region. The letters correspond with those on Fig. 101.

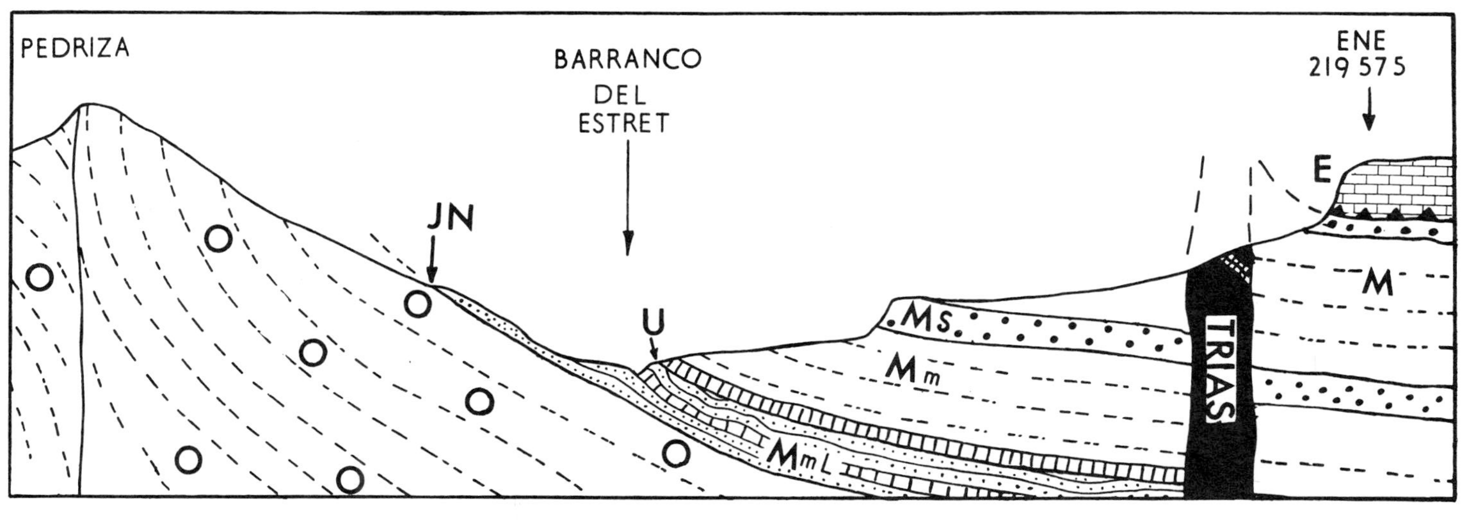

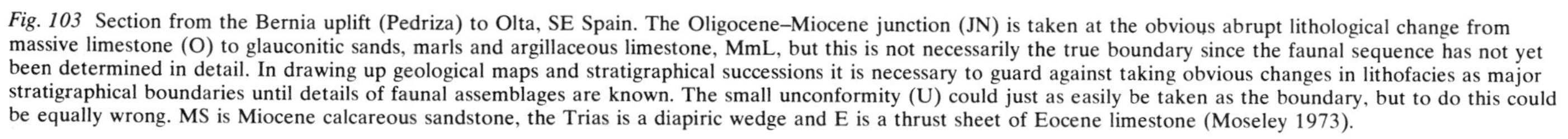

Fig. 103 Section from the Bernia uplift (Pedriza) to Olta, SE Spain. The Oligocene–Miocene junction (JN) is taken at the obvious abrupt lithological change from massive limestone (O) to glauconitic sands, marls and argillaceous limestone, MmL, but this is not necessarily the true boundary since the faunal sequence has not yet been determined in detail. In drawing up geological maps and stratigraphical successions it is necessary to guard against taking obvious changes in lithofacies as major stratigraphical boundaries until details of faunal assemblages are known. The small unconformity (U) could just as easily be taken as the boundary, but to do this could be equally wrong. MS is Miocene calcareous sandstone, the Trias is a diapiric wedge and E is a thrust sheet of Eocene limestone (Moseley 1973).

15 Upper Tertiary coastal sections in Cyprus: the Akrotiri cliffs

Cliffs sections are common in all parts of the world and can be recorded in the manner described here, regardless of rock type, structure or climate. Sea cliffs and mountain precipices can be treated in much the same way but details will depend on, for example, the availability of a boat for the former and access for the latter. The Akrotiri Peninsular ends in an E–W line of cliffs, which at their highest are about 60 m (Figs 104 and 105). Examination of the cliffs was primarily to determine whether any of the rocks would be suitable for construction and repair of the mole of a small harbour 200 m north of Cape Gata (Fig. 104 and Moseley 1976). The budget for this work was very small, and stone from established quarries elsewhere on Cyprus had proved too expensive.

The cliffs expose an excellent section of the Miocene Dhali Group; this is overlain by the Pliocene Nicosia Formation and Athalassa Conglomerate, and finally Quaternary calcrete and sand. Each unit is separated from the next by a small unconformity (Figs 105 and 106). The Dhali Group consists of chalk, chalk marl, shell marl and sand; the Nicosia Formation is mostly shell marl, shell sand and limestone; and the Athalassa Conglomerate also has some shell sands. The structure is simple, with the strata nearly horizontal.

Fieldwork was limited to one week, and with time at a premium and some 9 km of cliffs to examine it was necessary to base the survey on the simple but efficient photographic method referred to in part I. The framework of the survey was provided by a stereoline overlap of photographs taken from a fast motor launch about 400 m out to sea, an operation which took only 20 minutes. Photographs covered the cliff-line (Fig. 105) and extended further west. Using a stereoscope it was then possible to examine all parts of the section in detail and to make correlations with the ground observations (see below).

It so happened on this occasion that conclusions concerning the harbour stone and location of a quarry (the locality extended north from the cliff top between 1200 and 1300 m on Fig. 105) were straightforward and easily decided after one day, so the remaining six days were spent examining the sequence in a little more detail.

Understandably undergraduates may not have access to a fast motor launch and may flinch at the cost of so many photographs, but nevertheless this method could easily be applied during undergraduate projects and could be of great value, saving a great deal of time and making sections much more accurate. A rowing boat is always a possibility for sea cliffs; inland mountain cliffs can be photographed on foot. I took my photographs using the same principle as vertical photographs from an aircraft, the only differences being that the camera was horizontal and the overlap had to be rather more than the normal 60% of vertical aerial photographs, because of the greater 'relief' imposed by small bays and headlands.

The next stage of the survey was to select several parts of the cliff for more detailed investigation. These sections were then examined in the normal way. Measurements of cliff height (using a 50 m tape) were taken at several points so that the scale of the photographs could be calibrated; horizontal distances were taken from the excellent 1:10 000 maps of this area. A more rigorous survey would have required more accurate measurement of the horizontal distances. I made a special examination of lithologies and collected specimens. All the rocks are highly fossiliferous, and with more time it would have been possible to complete interesting faunal studies. Fig. 107 consists of sections compiled from the photoline overlap and from ground observations. Although the fieldwork lasted only one week, the follow-up laboratory work (identification of specimens, examination of the photographs, preparation of diagrams and writing the account), took rather longer; it is generally the time available in the field which is limited, and all ways in which this can be used with greater efficiency need to be explored.

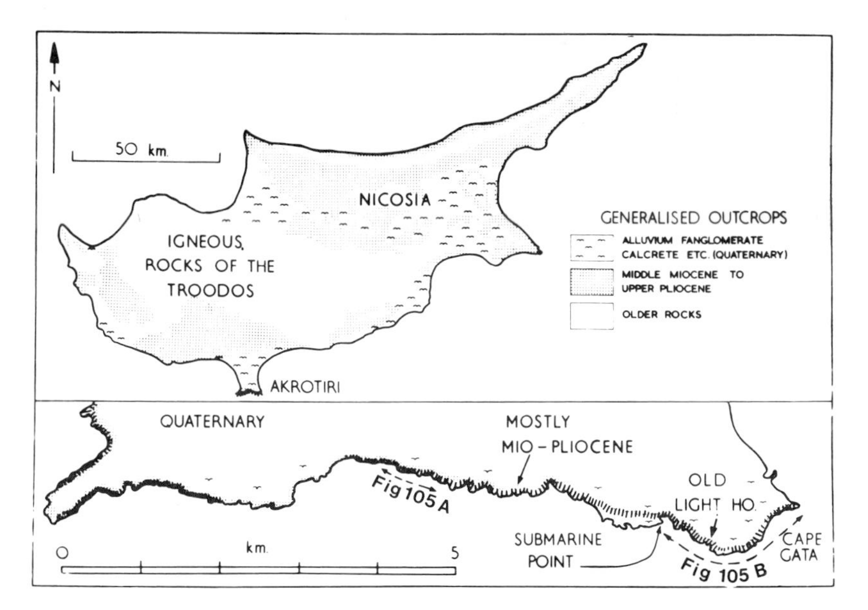

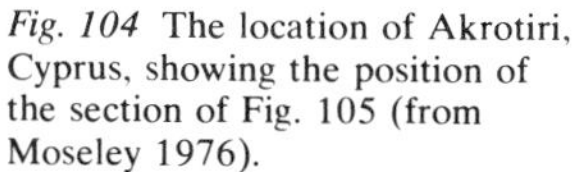
Fig. 104 The location of Akrotiri, Cyprus, showing the position of the section of Fig. 105 (from Moseley 1976).

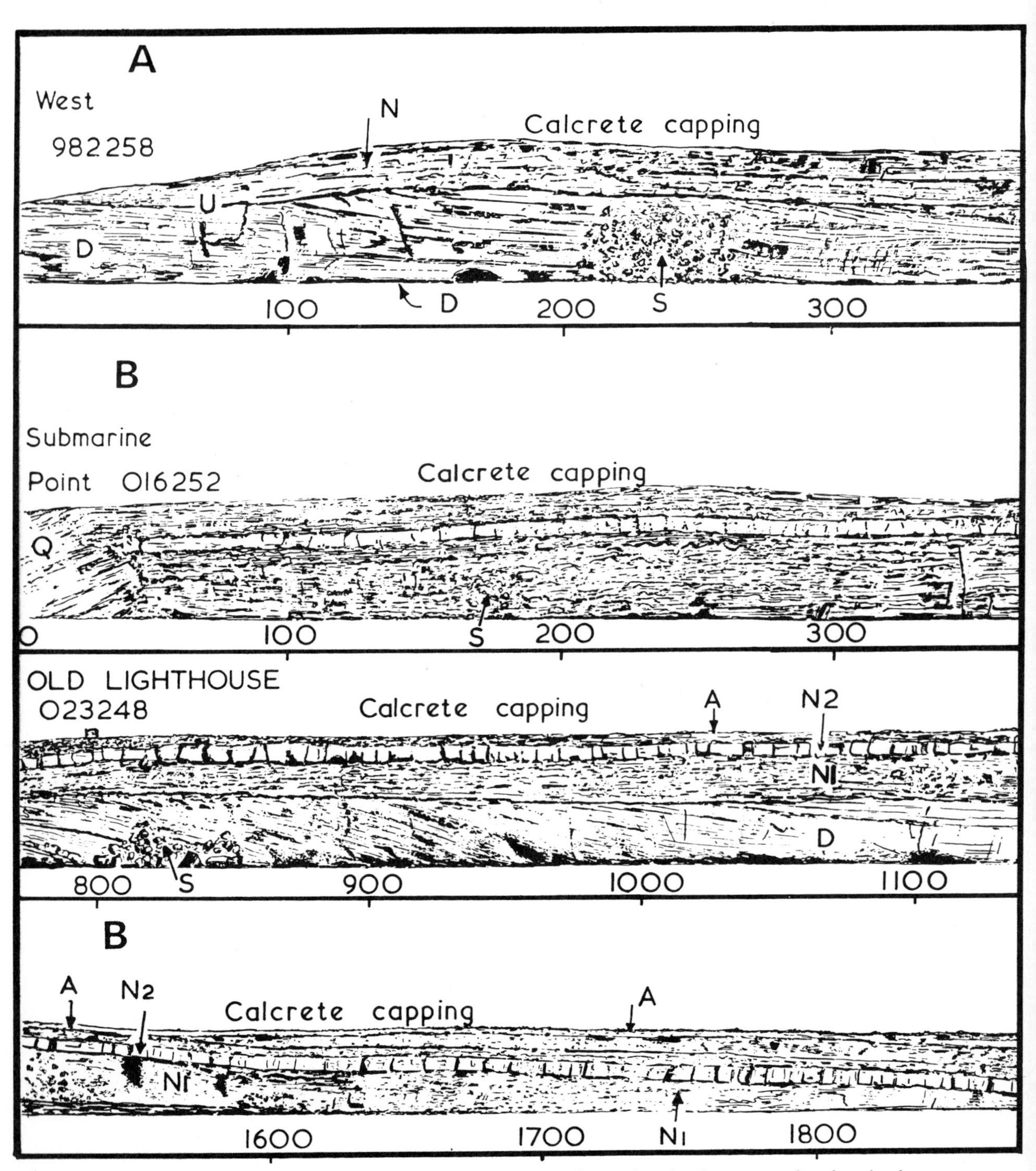

Fig. 105 A section of part of the Akrotiri cliffs, Cyprus, constructed from a stereo-line overlap taken from a motor launch, and scale corrected from cliff measurements (from Moseley 1976).

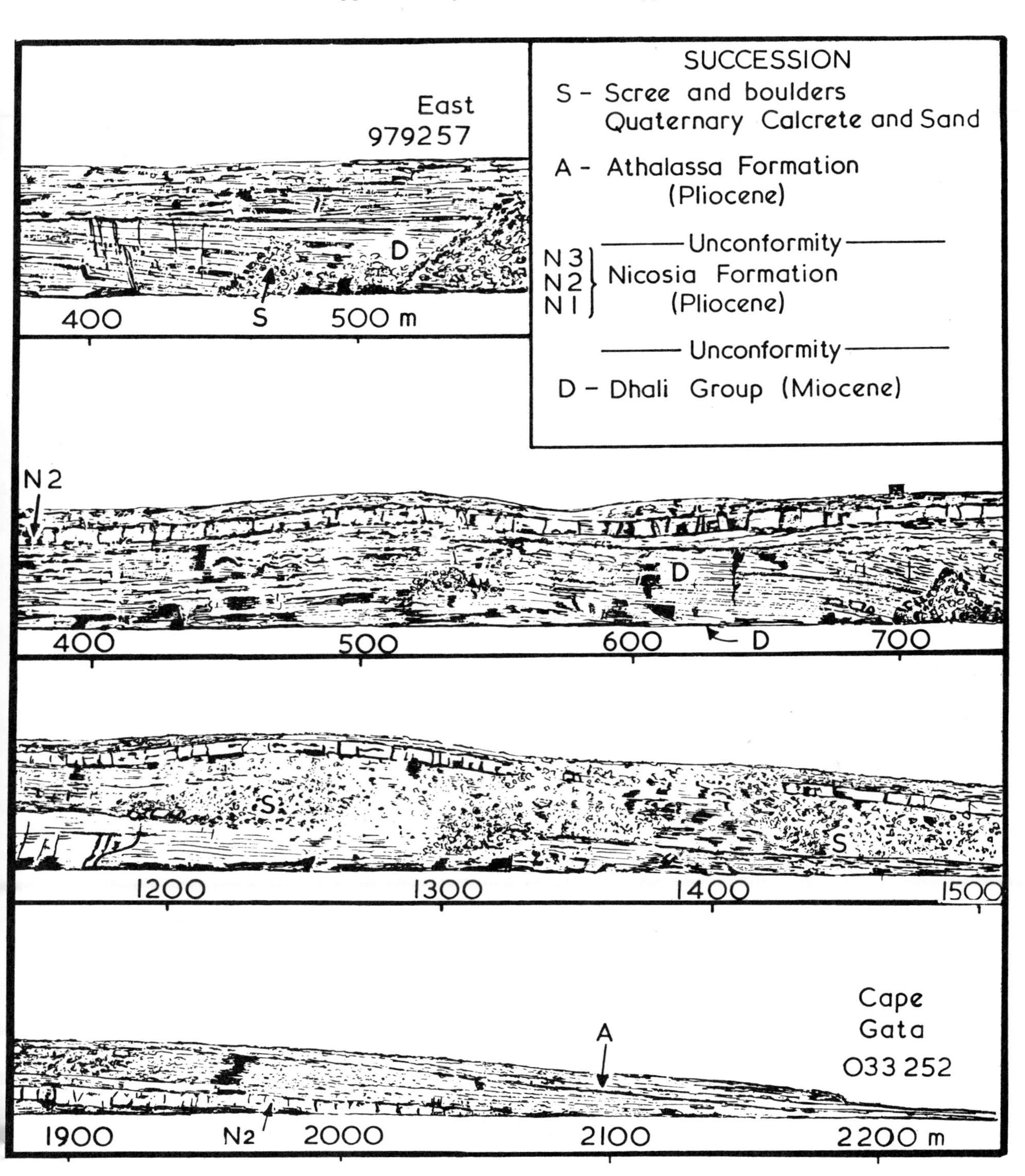
East
979257
D
S
400
500 m
SUCCESSION
S - Scree and boulders
Quaternary Calcrete and Sand
A - Athalassa Formation
(Pliocene)
Unconformity
N 3
N 2
N 1
Nicosia Formation
(Pliocene)
Unconformity
D - Dhali Group (Miocene)
N 2
D
400
500
600
D
700
S
S
1200
1300
1400
1500
Cape
Gata
033 252
A
N2
1900
2000
2100
2200 m

Q
A
N3
N2
N1

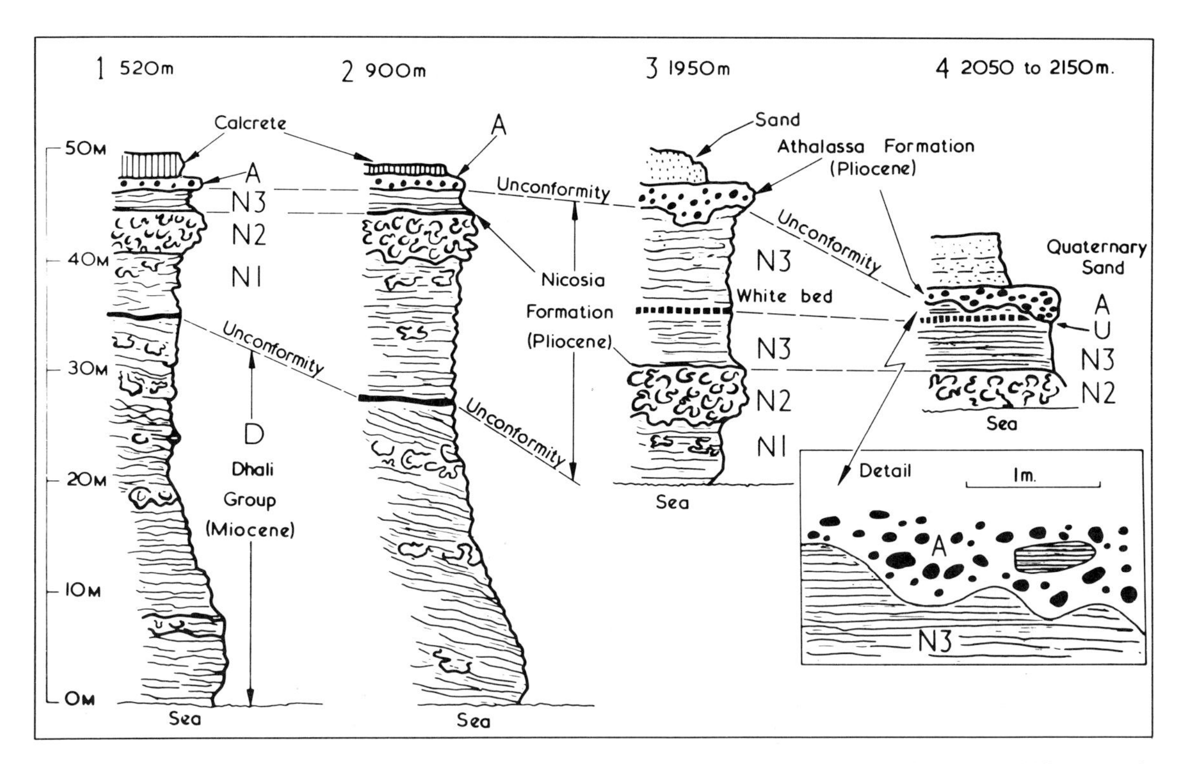

Fig. 107 Measured sections of the Akrotiri cliffs to supplement the photogeology of Fig. 105. The approximately parallel lines are soft marls and unconsolidated sands; the irregular ornament (e.g. N2) represents well-cemented shell limestone; the black spots are conglomerate and conglomeratic limestone; and there is a superficial cover of calcrete (havara) and sand. The final stage of a field investigation of this kind would be to make a careful collection from these sections, recording the fauna (and flora) and lithological details, metre by metre. (From Moseley 1976.)

◁ *Fig. 106 Above.* One photograph from the stereo-line overlap of the Akrotiri cliffs, Cyprus. Nicosia Beds (Pliocene) rest unconformably on Dhali Beds (Miocene). The position is at 900 m on Fig. 105. It will be appreciated that when cliffs are vertical as in this case, stereophotographs taken from some distance out to sea can reveal structures and sequences which a purely land-based survey may take a long time to determine.

Below. Detail of the sequence at the top of the cliff at 1100 m, Fig. 105. The location is shown on Fig. 104. Q, Quaternary calcrete (havara); A, Athalassa conglomeratic shell limestone; N3, loosely cemented shell sand and silt with irregular shell masses; N2, well-cemented shell limestone, about 3 m thick; N1, similar to N3. Photographed and measured sections of this type were used in conjunction with Fig. 105 to prepare sections such as those of Fig. 107. (From Moseley 1976.)

16 The Upper Cretaceous Ophiolite Complex of Masirah, Oman, SE Arabia

This survey was essentially reconnaissance in character, although several selected areas were mapped in detail. The methods used apply to regions not covered by large-scale topographical maps, and therefore would not be used in regions such as western Europe and the more highly populated areas of North America. There are, however, many regions where such maps do not exist; for example, vast areas of the Sahara Desert and the savannah lands further south, much of central Australia, parts of Brazil, Peru, Bolivia, India and even some parts of the deserts of the United States.

Masirah Island is 70 km long by 16 km wide and is about 20 km off the south-east coast of Arabia (Fig. 108). It is composed almost

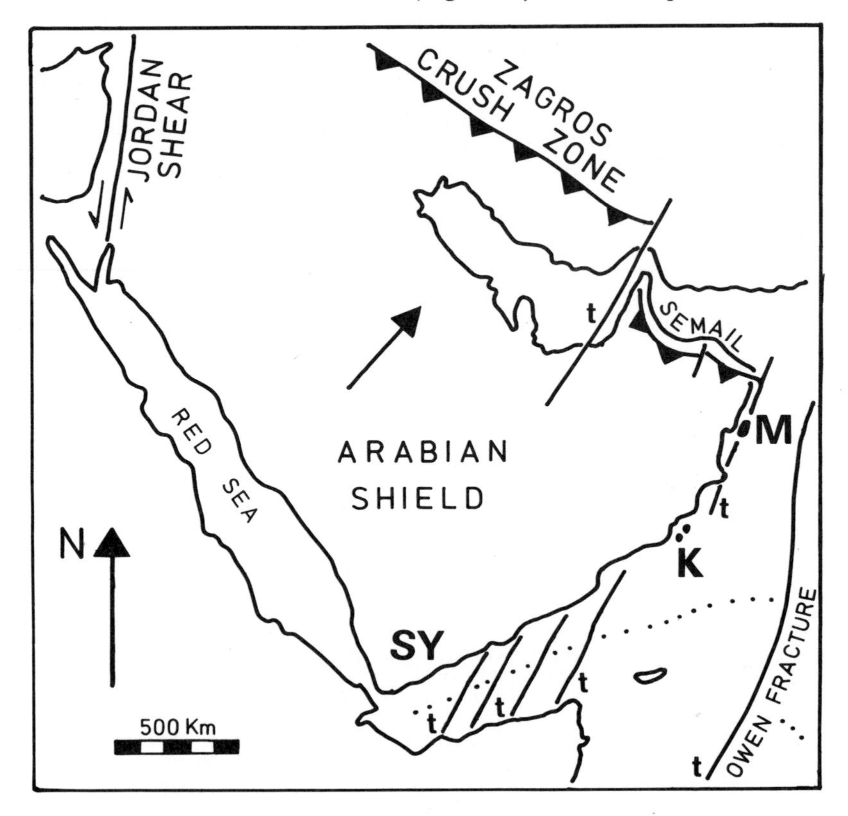

Fig. 108 Arabia and the position of Masirah (M) in its tectonic setting. K, Kuria Muria Isles (see Figs 13 and 20); SY, South Yemen (see chapter 17); t, transform fault. Modified from Gass and Gibson (1969).

entirely of an ophiolite complex of serpentinites, gabbroic intrusions, an extensive sheeted dyke complex, pillow lavas, red cherts, limestones and marls (Fig. 109). There is also a 5-km wide vertical mélange zone with blocks up to 2 km long. The objectives of our two expeditions (1976 and 1977) were to examine this complex and to establish its relationships to the Semail Ophiolite of the Oman Mountains, to the Arabian Continent and to the evolution of the Indian Ocean.

Surveys of this kind in fairly remote regions pose logistic as well as geological problems, and when the survey is organized by a university with limited funds, the whole burden of the logistics generally falls on the leader. This is frequently time consuming and frustrating since one is anxious to get on with the geology. It includes such items as permission from the government concerned, finance, travel to the region, supplies and transport within the area to be surveyed, availability of aerial photographs (maps are generally inadequate), and so on. It must be said, however, that the Masirah expeditions were fortunate in most of these respects. Air transport to the island from the United Kingdom was provided by the Royal Air Force, and they also airfreighted some 2 tons of rock specimens back to the UK, all without charge. This reduced the budget considerably; it is worth mentioning that government agencies, if approached in the proper way, are usually pleased to help with scientific expeditions. The RAF also provided a main base on the island (the Officers' Mess), stores, camping equipment and they arranged for a Land Rover, so the survey conditions could scarcely be described as rigorous. As a result our party of two was able to spend a total of 104 'man-days' of fieldwork out of a total of 110 man-days on the island. The total cost, including the processing of all the photographs, was £1300 (1976 prices), and this was covered by a grant from the Natural Environment Research Council.

Other factors contributing to the success of the survey were the low annual rainfall of less than 20 mm, which resulted, in this desert environment, of almost continuous exposure away from wadi alluvium, and the availability of stereo air photographs at scales of 1:12 000 and 1:80 000. There was also an excellent mosaic at 1:40 000 prepared by the RAF and a satellite photograph at 1:250 000. It is important to stress the value of these photographs. The smaller scales permit an overview of the whole region, the larger scales allow detail to be plotted. I wish to dispel a currently held view that satellite photographs will solve all problems. On the satellite photograph the whole island can be seen at a glance, but the detail is not there: the sheeted dyke swarm and the intrusive lenses of granite are visible only on the 1:12 000 scale. As a general guide a good photograph for a geological survey of this kind is one on which individual trees can easily be seen.

It will be evident that a survey of some 1000 km^2 in just over 100 man-days has to be of a reconnaissance nature. The island had been

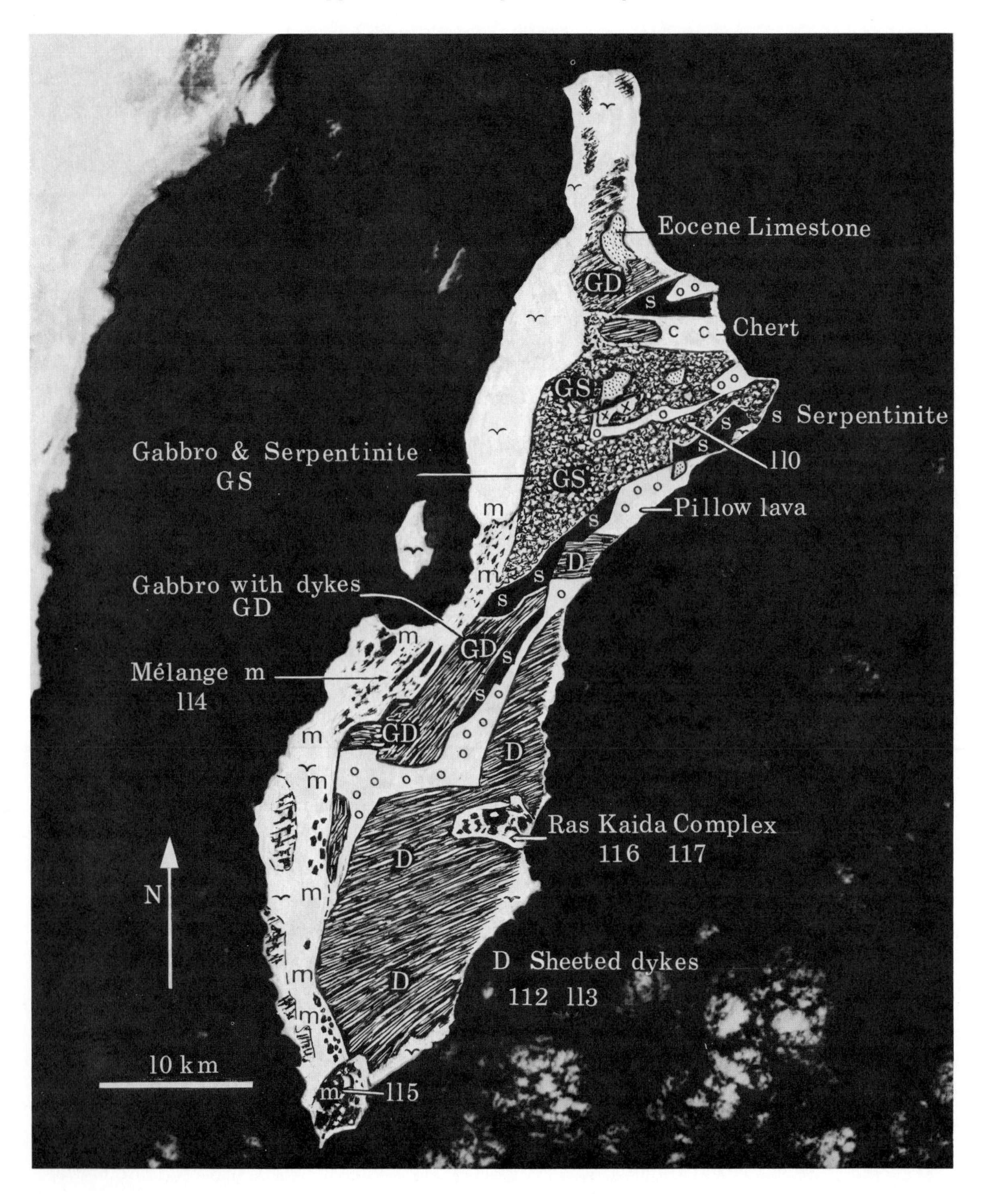

Eocene Limestone
GD
s
c c
Chert
GS
s Serpentinite
s
s
110
Gabbro & Serpentinite
GS
GS
Pillow lava
m
s
D
Gabbro with dykes
GD
m
s
s
GD
Mélange m
114
s
m
GD
D
m
m
Ras Kaida Complex
116 117
D
N
m
D Sheeted dykes
112 113
m
D
m
10 km
m
115

briefly visited previously by a few geologists (Glennie *et al.* 1974; Lees 1928; Moseley 1969a) but no details were known. The methods of tackling a problem where only reconnaissance is possible yet details are required are of importance.

Initial preparation

Vitally important work had to be done before leaving the United Kingdom. Since it was known that there was an ophiolite complex on Masirah, it was obviously necessary to be familiar not only with the literature of nearby parts of Arabia and the Indian Ocean, but also with relevant problems in other parts of the world. Geological surveys of this kind need to be multidisciplinary since the field survey is followed by geochemistry, geophysics, the preparation of a geological map and some palaeontology, and so all the relevant literature had to be read beforehand. Preparation of a photogeological map was also of great importance. The 1:12 000 stereophotographs made such a map possible; although the meanings of some boundaries, lineaments and rock types were not known, the map proved to be a good approximation. It made it possible to decide upon the most profitable geological traverses, those areas where exposure was less satisfactory, those where there was uniform geology and, most important, those well-exposed regions that appeared critical to the understanding of the complex and which should therefore be examined in greater detail. The field relations of the granite intrusives, the nature of the mélange zone, the relation between pillow lavas, limestones and cherts and details of some plutonic complexes come under this last heading. A reconnaissance survey of this kind thus has to divide time between the small important areas, where survey is concentrated, and the remainder, covered by reconnaissance traverses.

Survey methods

In the field aerial photographs with a pocket stereoscope were used at all times (see chapter 4). This was absolutely essential both for locating position and interpreting geology. During the early part of the survey there was a general reconnaissance both of the areas previously

◁ *Fig. 109* Geological map of Masirah, Oman drawn on a satellite photograph. The locations of the text figures that follow are indicated (110 etc.). The serpentinite (S) is serpentinized hartzburgite with minor gabbro veins and anorthositic lenses and is considered to be part of the mantle (Abbotts 1979). Gabbro and serpentinite (GS) includes serpentinized troctolite and picrite of the lower crust, the Ras Kaida intrusion being part of this complexity. GD includes gabbro with dolerite dykes, in places not unlike (D) (dykes with gabbro screens), but there are also gabbro intrusions discordant to the dykes. This complex is in general at a higher structural level than GS. The sheeted dykes (D) at a still higher structural level vary from 100% dolerite dykes to dykes with substantial proportions of gabbro screen (up to 70%). The pillow lavas (interbedded with marl and limestone) and chert are in tectonic zones, and generally have steep dips. The mélange (m) truncates the ophiolite on its western side, and forms the outcrop of the Masirah Fault (Fig. 108). X (north centre) represents the outcrop of potash-rich granite considered to be a continental crust contaminant (Abbotts 1978). An upper age limit to the ophiolite is given by the unconformable Lower Tertiary limestones (dots) (Fig. 110).

selected for traverses and those selected for more detailed work. These were refined as the work proceeded until it was known which parts of the island could be explained by reconnaissance traverses and which parts would require the more detailed work. The objectives were now delineated as follows:

1. To ensure the traverses crossed major structures seen on the photographs.
2. To map the critical areas in as much detail as time permitted (1:12 000 scale usually).
3. To collect numerous accurately located specimens. It was necessary that they be large enough for geochemical work (say 8 cm^3). A total of 1500 specimens were collected of which a number of dyke rocks were oriented for future palaeomagnetic work.
4. To supplement field sketches with photographs; for example, Figs 110, 111 and 117. (It is impossible to memorize complex exposure details—words in a notebook do not suffice—and although accurate field sketches are necessary, they cannot tell all.) In many cases the cost of the stereophotographs and stereopanoramas (chapter 8) determined that most of them had to be black and white rather than colour; in all 600 black-and-white and 200 colour photographs were taken.

Reconnaissance traverses

Two methods were used according to the terrain. The most convenient was when a wadi could be followed by Land Rover and stops made at appropriate geological locations. Where the terrain was rough, the traverse was on foot. In the latter case the weight of the rucksack was a problem. At the beginning there would be a 1-gallon water bottle, cameras, lenses, large hammer, map case, etc. The water steadily decreased but the specimens increased, so that by the end of the day the rucksack was unlikely to weigh less than 50 pounds—nothing to a Himalayan porter, but considerable for an ageing geologist. There were days when RAF personnel asked to come out with us out of general interest, and were able to add to their physical fitness by taking some of the specimens.

A one-day Land Rover traverse (example)

This particular traverse in the south of the island followed two major wadi systems from the east to the west coasts, crossing a large section of

Fig. 110 A northward view towards the Eocene limestone outlier of Jabal Hamra (E), Masirah ▷ shows the nature of the terrain. This is the highest part of the island, about 800 feet above sea level, but in spite of the low altitude the ground is divided into numerous steep craggy ridges and valleys, so that the impression is of a truly mountainous region. The ophiolite structure in this area is complex, with a steeply inclined strip of marl–limestone interbeds (m) and basaltic pillow lava (b) (ocean floor formations), tectonically emplaced within lower crustal ultramafic rocks (mostly) serpentinite, troctolite and gabbro (u). There are gravel terraces in the foreground (t).

the dyke swarm. Frequent stops were made to measure the orientations and thickness of dykes and the proportions of dyke to gabbro screen (Figs 112 and 113). Specimens were collected for geochemistry and palaeomagnetism (the latter oriented), and a continuous check was kept on the aerial photograph positions. It was soon apparent that a combination of this one section and aerial photograph interpretation would make possible a satisfactory map of a wide area and therefore detailed mapping was not undertaken in this region. Fig. 113 is a photogeological map prepared by enhancement of photocopies of the aerial photographs (see also Figs 48 and 116).

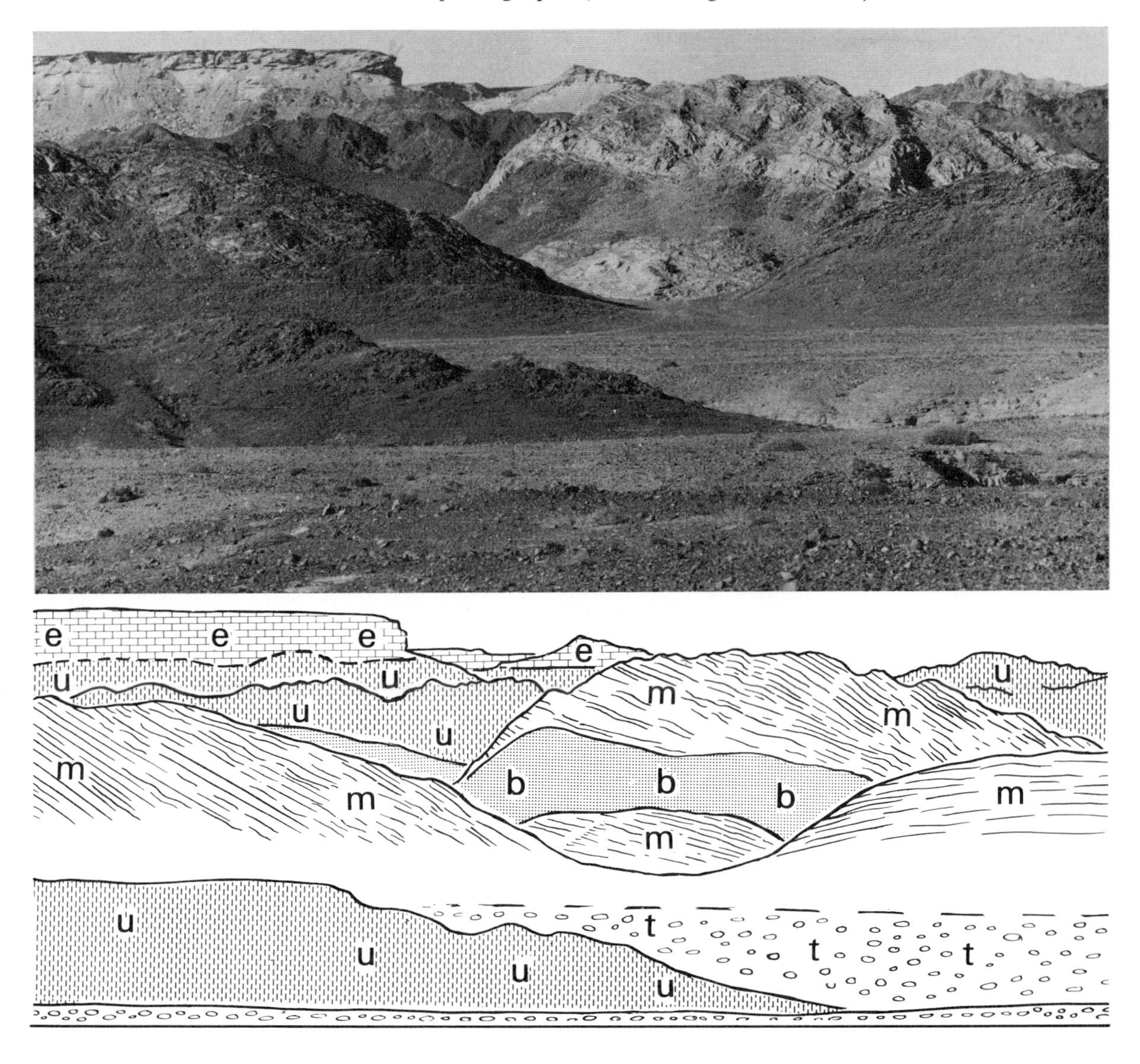

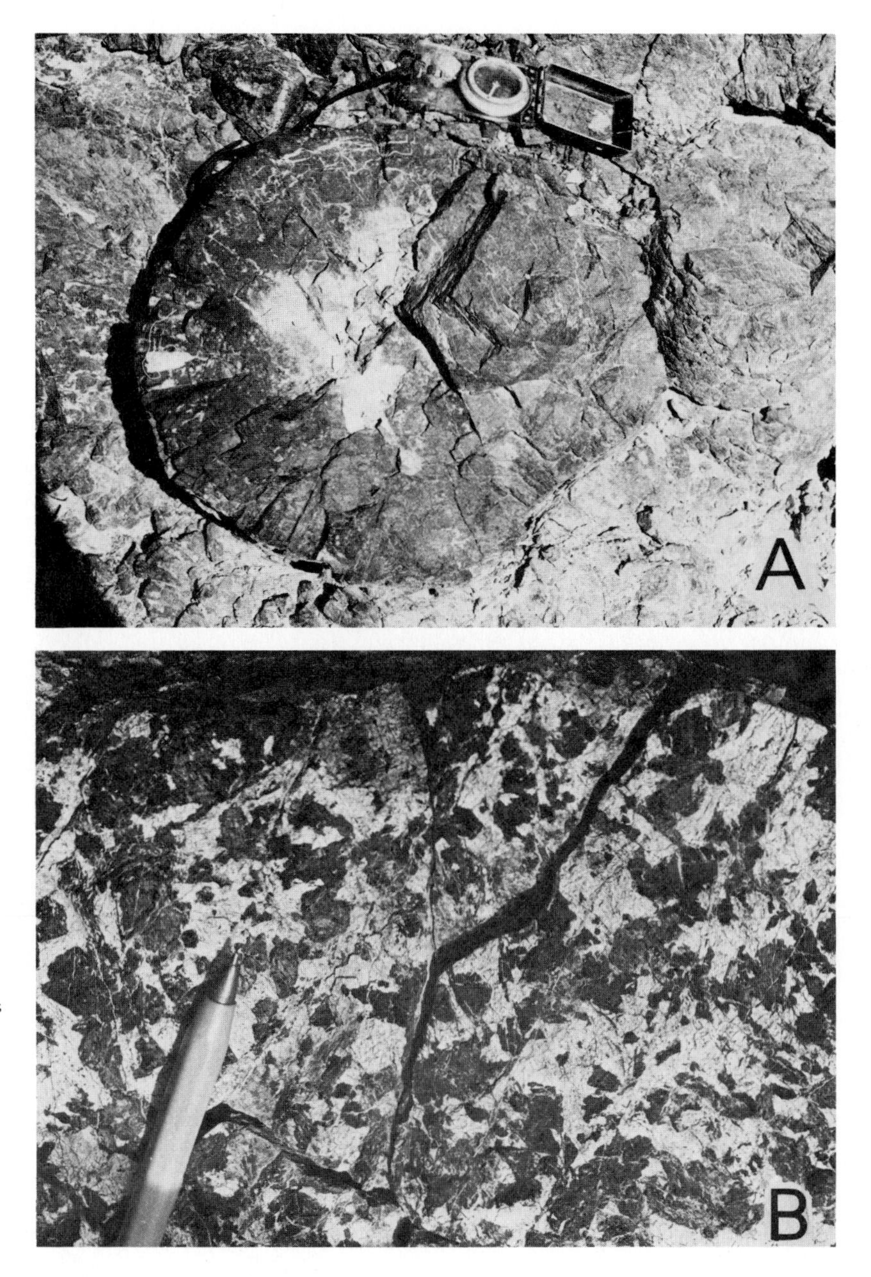

Fig. 111 **A**, A single pillow from the pillow lava sequence in Masirah. Numerous specimens, including a whole pillow were brought back for petrological and geochemical examination (Abbotts 1979). **B**, Coarse-grained gabbro grading to gabbro pegmatite. These rocks are characteristic of the lower part of the cumulate sequence forming the lower part of the crust. One difficulty in the field was the collection of fresh, reasonably large samples, since these rocks almost always disintegrated into individual crystals on being hammered.

△
Fig. 112 Part of the sheeted dyke complex of south Masirah. In this exposure dolerite dykes and gabbro screens are each about 50%. The drawing was an enhancement of a photocopy as described in Part I (see Fig. 43). The appearance of outcrops such as this on aerial photographs is shown on Fig. 113.

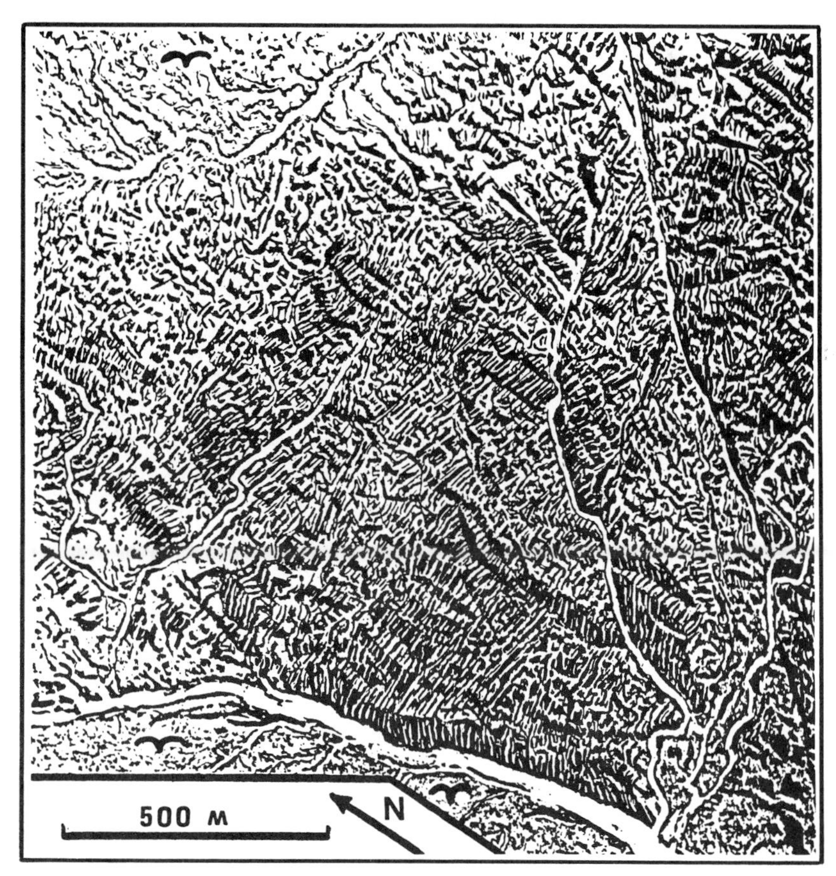

Fig. 113 Photogeological map ▷ (the initial interpretation) of part of the sheeted dyke complex of south Masirah. The topography consists of numerous sharp ridges about 50 m high which trend approximately N–S. The morning sun casts shadows on their west-facing slopes. Most of the dykes have E–W trends and steep southerly dips (seen from the way they cross the ridges, Fig. 4). Maps of this type make it possible to plan traverses; in this case the logical traverse was along the wadi on the western side of the map, an easy route at right angles to the dyke trend with the well-exposed ridge rising to the east. It is interesting to observe that the dykes are clearly seen on the 1:10 000 photographs, but only the N–S ridges are visible on smaller scale photographs (1:60 000 etc.). Had large-scale photographs not been available, photogeological interpretation would have been inadequate. The map was drawn under a stereoscope by enhancement of an aerial photograph (photocopy).

Detailed mapping

Several more important areas were mapped on a scale of 1:12 000 as it seemed probable that a thorough understanding of these areas would be a key to understanding the whole ophiolite complex. Two such areas were as follows:

Masirah Mélange. This is a zone of mega-breccia, 5 km wide, which runs down the west side of the island and contains blocks, from 2 km to 2 m in diameter, of all the rocks within the ophiolite, that is serpentinite, gabbro, dolerite, pillow lava, chert and limestone. It has been interpreted (Moseley and Abbotts 1979) as a section of a former transform fault, and since such structures are rarely seen on land areas where they can be examined in detail, it was important to take every advantage of this opportunity. The total exposed length of the Mélange Zone is 40 km, so it was clearly not possible to map all of it in detail. The best exposed parts were selected, some of which are illustrated by Figs 114 and 115. The remainder was interpreted from aerial photographs supported by traverses.

Ras Kaida Complex. This complex is a plutonic intrusion of gabbro (much of it layered) and ultramafics (serpentinite and serpentinized troctolite). The latter occur as megaxenoliths up to 1 km diameter within the gabbro. It penetrates to a high crustal level in the sheeted dyke complex, and it was immediately apparent from the aerial photographs that it was a region of great interest and one to be examined in detail. It has now been interpreted as an off-axis intrusion (Abbotts 1979).

The survey method was to take the Land Rover to central positions along the wadis, and then to survey on foot as indicated on Fig. 116. The aerial photographs gave extremely reliable indications of geology hereabouts. The map (Fig. 116) is an enhancement of an aerial photograph mosaic, which includes results of the ground observations and ground photographs (Fig. 117). Detailed identification of the rock types had to await petrological and geochemical studies during the next year at Birmingham.

A conclusion

Completion of the field survey was only a beginning to the understanding of the geology of Masirah. As with all postgraduate fieldwork and to a lesser extent with undergraduate projects the follow-up laboratory work is most important. Most of this work was completed by I. L. Abbotts, the other half of the survey team, for his PhD degree, and was largely in three parts:

1. Petrology of thin sections. We had 1500 specimens but only a proportion were sectioned.

2. Geochemistry, a tool which is becoming increasingly important on all types of rocks. This was mainly X-ray fluorescence analysis at Birmingham, but it was supported by election probe work at Leicester University.
3. Preparation of geological maps of the island from aerial photographs and ground observations, at a scale of 1:25 000.

Other aspects of the geology—such as the palaeomagnetism of the oriented dyke rocks and determination of the lower Tertiary limestones which unconformably overlie the ophiolite—were dealt with by specialists.

It is clearly of the greatest importance that work of this kind is published as soon as possible, and this raises an important point where PhD theses are concerned. Many of these theses are detailed accounts, many hundreds of pages long, containing tables of analyses, numerous diagrams and photographs, all interleaved with, and essential to the reading of the text. Such a thesis is entirely unsuitable for publication since few journals will take papers of more than 20 pages, and the thesis will have to be entirely rewritten, perhaps as three or four independent papers, for this purpose. I suggest that this type of thesis layout is not only a waste of effort but it frequently results in non-publication. The person concerned takes up lucrative new employment which requires full-time concentration, and the rewrite is never completed. There are many important PhD theses languishing in university libraries, unknown to the world. This should not be permitted. The theses should be organized from the outset for publication with each major chapter complete in itself, ready for despatch to a journal. The mass of detail unlikely to be published (for example, X-ray fluorescence methods and detailed tables) can then be included as appendices for reference.

In the case of Masirah, Dr Abbotts completed his PhD by the end of 1978; by late 1979 three chapters were already published and others were in press. In another case Dr D. E. Roberts, within a few years, had published six papers from his thesis on structures of the Skiddaw Slates (Roberts 1973).

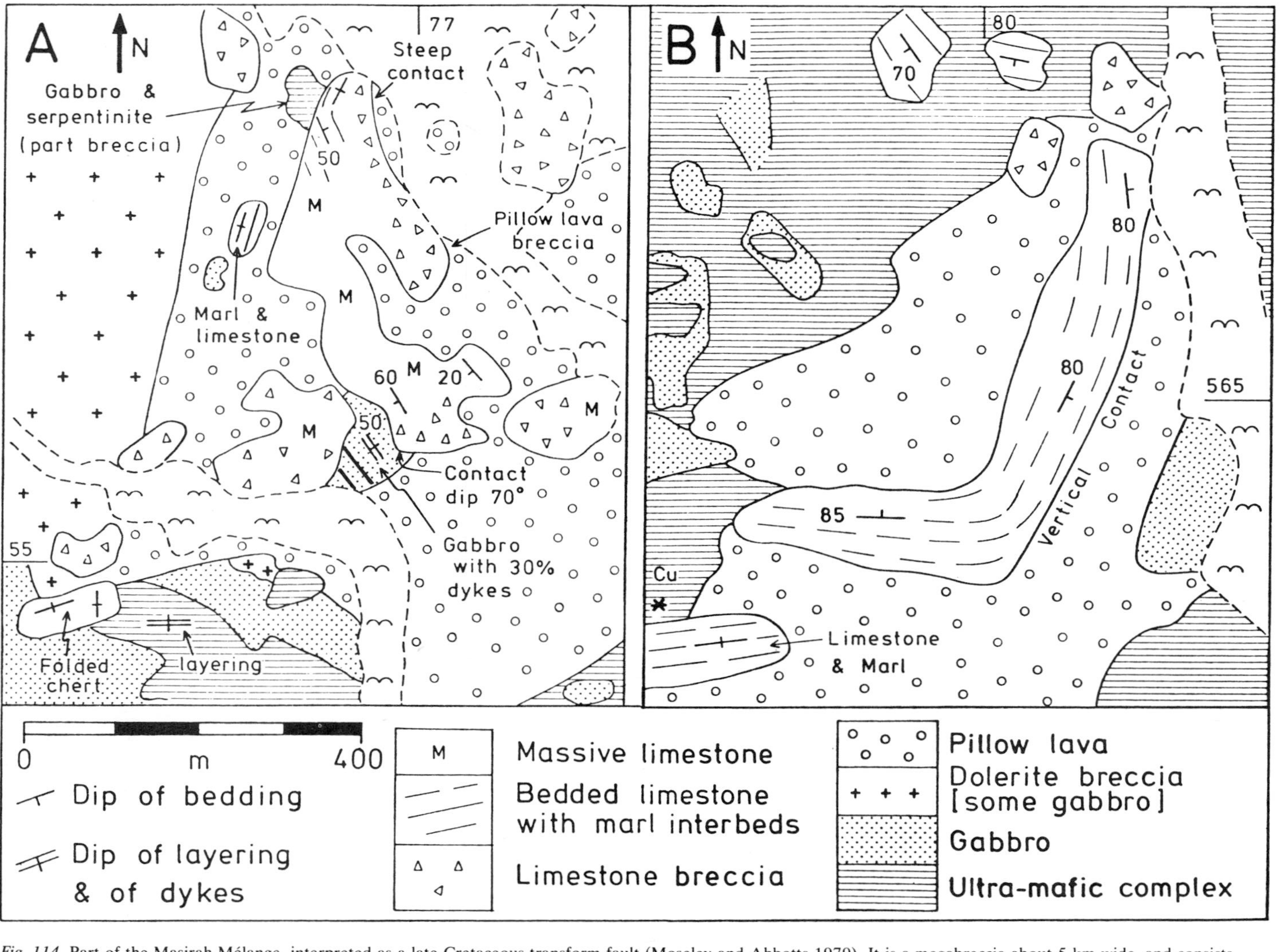

Fig. 114 Part of the Masirah Mélange, interpreted as a late Cretaceous transform fault (Moseley and Abbotts 1979). It is a megabreccia about 5 km wide, and consists of blocks of all the ophiolite lithologies, which range from several kilometres to several centimetres in diameter. The areas that seemed to be the most interesting and best exposed were selected following photogeology and ground traverses, and these were then mapped in detail. An area such as this one is a key to the understanding of the whole of the mélange zone. (© Geological Society of London.)

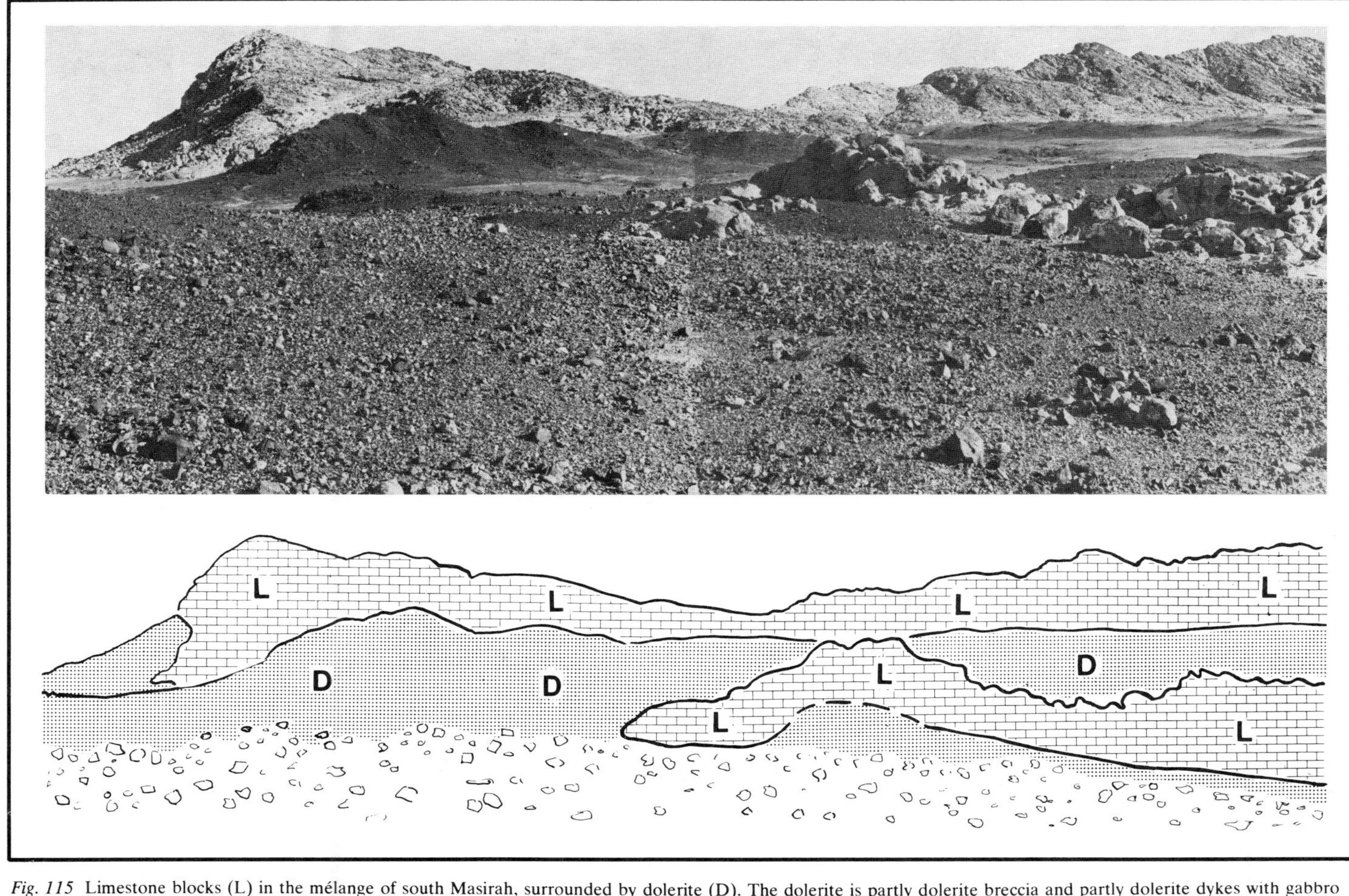

Fig. 115 Limestone blocks (L) in the mélange of south Masirah, surrounded by dolerite (D). The dolerite is partly dolerite breccia and partly dolerite dykes with gabbro screens, and like the limestone is in the form of blocks incorporated in the mélange. It is 1.5 km from the left (north) to the right of the photograph.

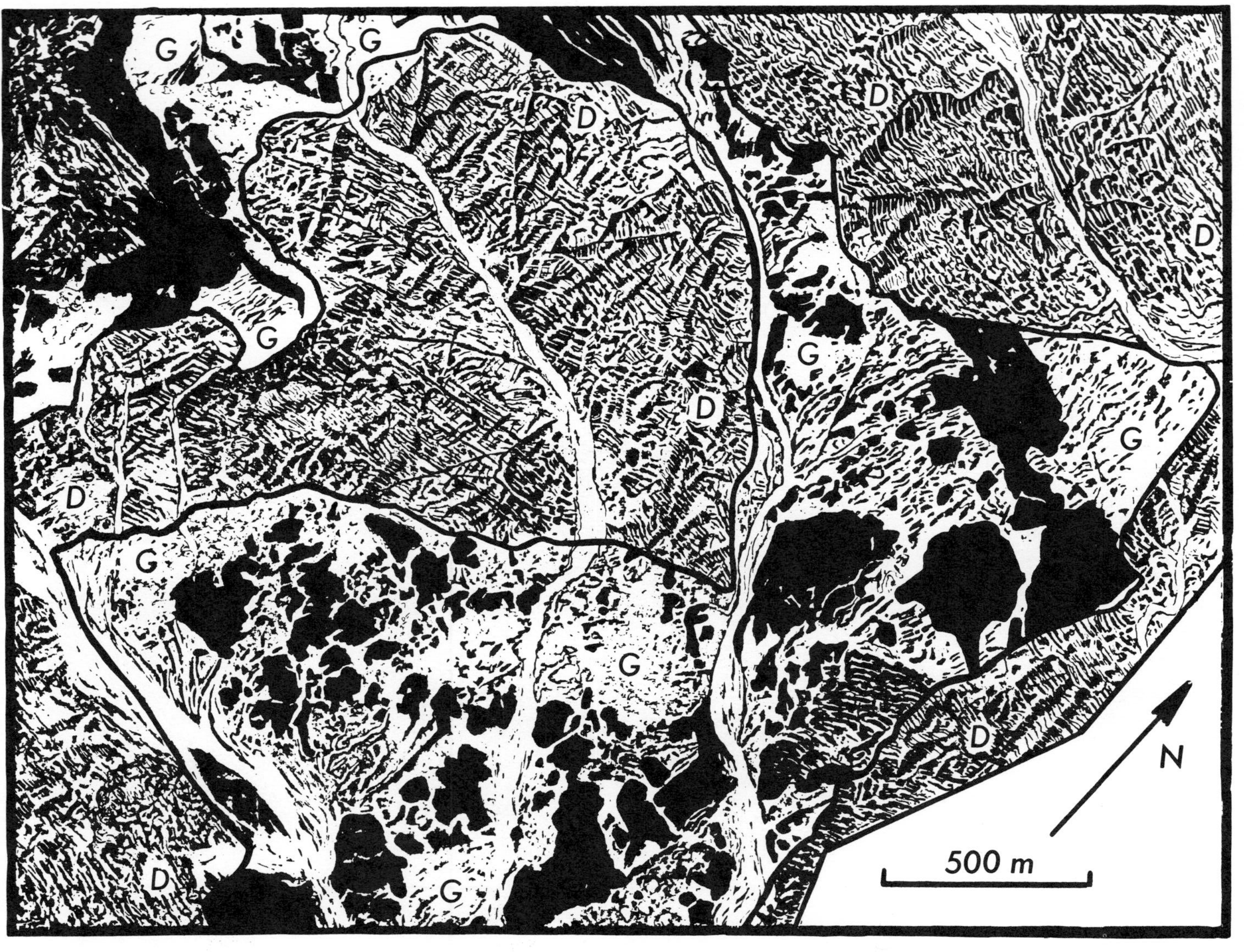

Fig. 116 Photogeological map of part of the Ras Kaida intrusive complex, Masirah. G, Gabbro, with layering visible in places (areas with black streaks); black, serpentinite (serpentinized periodotite and troctolite) which occurs as macroxenoliths in the gabbro; D, sheeted dykes, either vertical or highly inclined (determined from the way they cross the north–south ridges, see Fig. 113). The narrow wadis between the ridges are mostly orientated along master joints. The region is traversed by major wadis, the field procedure being to drive along these to a point selected from the photogeological map, and then to traverse on foot. The map was drawn by enhancement of an aerial photograph mosaic (photocopies), and was subsequently simplified so that field data could be added.

Fig. 117 Part of the Ras Kaida complex, Masirah showing extensive outcrops of gabbro (grey) and serpentinite (black). Numerous photographs of this and other areas were essential since it was impossible to remember or accurately to sketch details of all outcrops.

17 Hydrogeological reconnaissance surveys in South Yemen, SW Arabia

Most of the comments at the beginning of chapter 16 apply to the South Yemen surveys, the difference being that the objectives in this case were hydrogeological. The methods used and the general conclusion would be equally valid in other desert regions of the world, for example the Simpson Desert of Australia, Thar Desert of India, the Atacama Desert of Chile or the Mojave Desert of the United States. The geology of South Yemen has most recently been described by Greenwood (1966) and Greenwood and Bleakley (1967).

These hydrogeological surveys were organized by the Royal Engineers and conducted at different periods over the three years 1965–1967 (Fig. 118). For any reconnaissance survey logistic support is most important. The optimum conditions are realized when surveys (generally with economic incentive) are backed by large organizations and the initial time-consuming administrative work is taken from the geologist, who can then proceed with his real job. In such conditions a great deal of geological work can be accomplished very quickly. On the opposite side of the coin, however, is the unfortunate fact that many expeditions are insufficiently funded. This particularly applies to expeditions of a fundamental scientific nature where there is no immediate economic gain; in these cases a few highly qualified expedition members may spend a disproportionate amount of time on relatively unproductive administrative work.

The Arabian surveys come into the first of the above categories and were blessed with an enlightened army command which provided all the aerial photographs required at the wave of a wand, not to mention comprehensive helicopter and ground vehicle transport, and the vitally important army escorts into remote and politically difficult terrains. In one case the whole operation from Britain to active survey in the desert was mounted in less than a week. The surveys were simple in conception, the requirement being to provide water supplies for army camps and for towns and villages, with a long-term but subsidiary view towards irrigation. For the most part the operating procedures were divided into six categories: (1) an initial photogeological investigation; (2) helicopter traverses; (3) reconnaissance by ground vehicle; (4) traverses on foot; (5) geophysical survey; (6) preparation of report.

The general requirement for each of the three surveys was to complete the fieldwork within one month and to make immediate recommendations that would enable well drillers to begin their programme. There was no time for careful consideration of all possible lines of

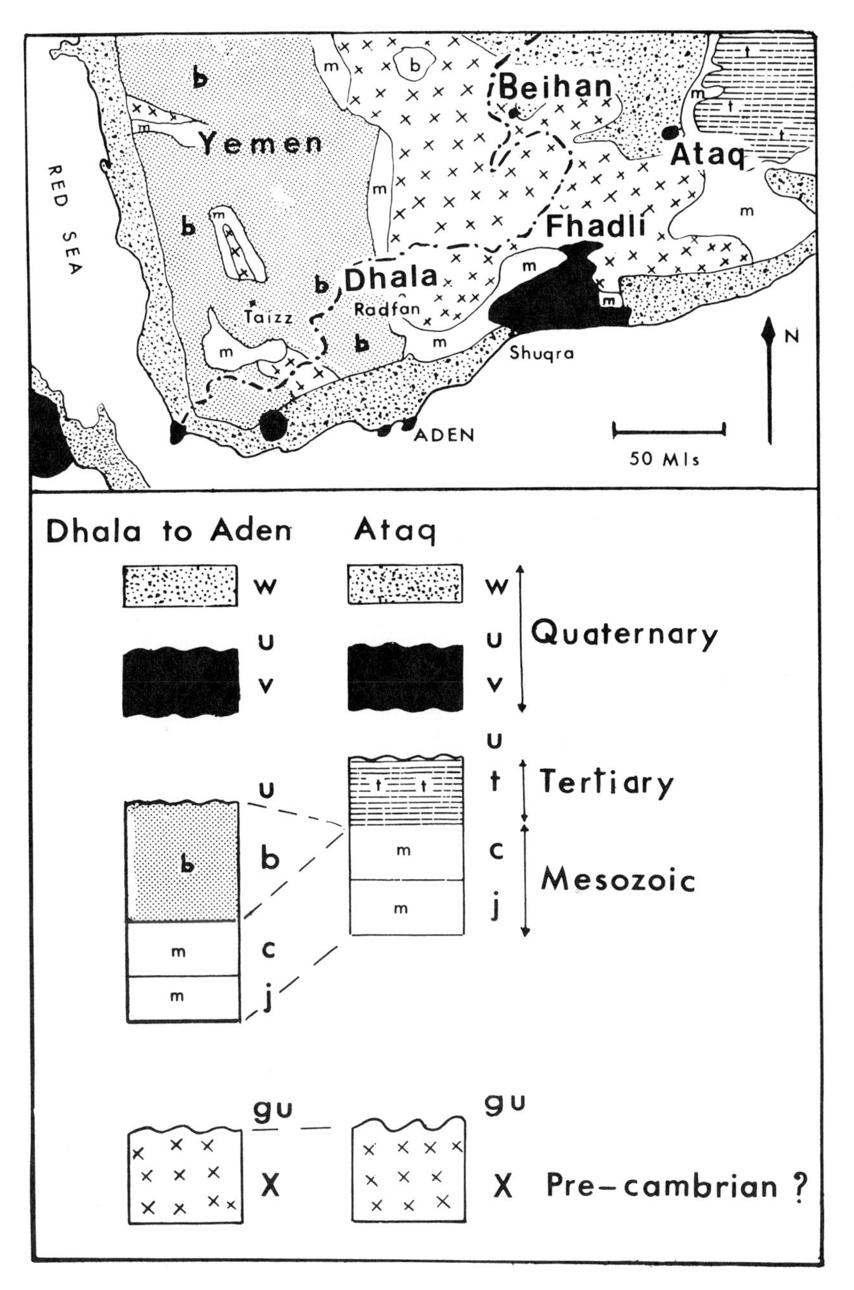

Fig. 118 Geological map and rock successions in SW Arabia . W, Wadi, dune and coastal fan sands and gravel; v, Recent volcanic rocks; t, Lower Tertiary limestones; b, Tertiary volcanic rocks (basalt); c, Cretaceous sandstones; j, Jurassic limestones; x, basement, mostly Precambrian (granite, gneiss etc.); u, unconformity; gu, great unconformity. The larger print refers to regions mentioned in the chapter and on Figs 35 and 140. (After Moseley 1966.)

evidence and a decision about exact locations of borehole sites had to be decided quickly: indeed several of the reports were written on the aircraft returning to UK. Nor were there facilities for geophysical traverses; which was unfortunate since several awkward problems could have been solved by seismic and resistivity survey.

The actual fieldwork was by no means straightforward and a number of difficulties, not all of them geological, quickly became apparent. The region was entirely undeveloped hydrogeologically, practically all the wells had been hand dug, many to depths exceeding 100 m, and showed exquisite workmanship, but most were within the range of 10 to 20 m. The successful wells had largely been developed by trial and error plus local knowledge over centuries, and it would be wrong to dismiss this factor in any assessment. However, there were sites which relied on superstition for their location. For example, one local untrained hydrogeologist was known as the 'dreamer' because of his method, but his success rate was not high. There were, of course, hazards in conducting a survey which required the presence of an army escort at all times, especially during the shooting season. The local sport was quite definitely shooting (there being no football), and it was generally conceded that a British party enjoyed the game of being hunted, much as foxes do in Britain. Once, after a survey had been completed and the most favourable locations for boreholes had been decided, a Sheik, delighted at the prospect of a borehole into a productive aquifer, nevertheless insisted on it being sited inside his fort. On another occasion an army commander required all water wells to be within the camp perimeter. The reasons become apparent when one considers the problems of well protection. In the Radfan Mountains there was one very good well into faulted sandstone about 600 m outside the perimeter of the army camp and dissident tribesmen made a commendable attempt to blow it up, in spite of covering machine guns.

Much of the vehicle travel is conveniently along the dry wadis, but this can also have its problems. There may be unexpected flash floods—perhaps a 5-m high wall of water rushing along the wadi in response to a thunderstorm 100 km away. Flash floods have been known to transport vehicles 10 km or more, usually in the wrong direction. Finally, there may be drilling problems at a borehole site. One such in Fahdli (Moseley 1971b) penetrated highly fissured limestones underlain by impermeable basement, and had a good chance of success. However, a drilling rig requires water for its operation, and in this case there was only one small (400-gallon) bowser (water truck) which had to transport water across the desert from the nearest well 16 km away. The limestone was so permeable that the 400 gallons was absorbed in a few minutes, and the problem was how to get the borehole completed with such limited resources.

Initial photogeological survey

It is tempting to refer to this side of the work as photo-hydrogeology, and it is with great satisfaction that one can record the accuracy with which preliminary hydrogeological assessments can be made from aerial photographs in many of these desert regions. For example, an initial 'photo-hydrogeological' report for Masirah Island, Oman (chapter 16) differed scarcely at all from one subsequently produced after geological reconnaissance which included an examination of existing wells; the same is true for several areas in South Yemen. The photogeological stage of the investigation essentially involves the production of a photogeological reconnaissance map, already discussed in chapter 3 (page 10), and illustrated by Figs 6 and 35. These maps permit estimation of the outcrop and dip of permeable and impermeable rocks, and of the extent and thickness of superficial sands and gravels. It is possible to calculate depths to aquifers and to locate suitable points for boreholes. Thicknesses of superficial deposits and borehole positions can also be suggested, although usually with greater uncertainty because of the high degree of variability shown by these deposits in desert regions. These maps are subsequently modified following field reconnaissance (Figs 119 and 121; see also Moseley 1971a).

Helicopter traverses

Use of a helicopter enables a geologist to obtain a bird's eye view of the ground at a closer range than that provided by aerial photographs. It is also possible to land and collect specimens from critical localities which would otherwise be comparatively inaccessible and perhaps not contemplated during a reconnaissance ground survey. Such localities include sharp mountain tops, remote gorges or high escarpments peripheral to the region of study. Visits to locations of this kind during the surveys of South Yemen included the steep south-westerly part of Fahdli (Moseley 1971a, fig. 2), the western and northern boundaries of Musaymir (Moseley 1969b) and the basement mountains flanking the Wadi Beihan (Moseley 1971b). A camera should obviously be taken and it should be continuously in use to ensure that a permanent record is available. It is in fact preferable to have two observers, one to operate the camera and the other to take notes integral to the camera shots. This method is not, of course, the only one which can be used during helicopter surveys. There are occasions when helicopter and ground surveys are combined, and the party is dropped at a selected point in the morning and picked up again later in the day.

The Fadhli survey, the object of which was to locate wells for villages, used both photogeological methods and helicopter traverses.

It was necessary here to examine a 30-km wide arid alluvial plain, surrounded and crossed by rough mountains and escarpments of Precambrian metamorphics, Jurassic limestones and recent volcanics. The topographical maps were crude and there were no geological maps; most of the interpretation was photogeological, but confirmatory ground checks were necessary. The first helicopter traverse covered the entire region of about 100 km^2 in one day, landing periodically to collect specimens. There were no roads and to cover the same region using a vehicle and on foot would certainly have taken more than a week. This traverse helped to decide those smaller areas which required more detailed study; thereafter the helicopter transported the party from a comfortable and secure base at Lawdar to a convenient starting point for survey on foot, with a pick up arranged for later in the day. Similar methods were used when surveying the Musaymir region (Moseley 1969b), a volcanic region with extremely rough terrain. In one case it was possible to examine for the first time the 1700 m spire of Jabal Warwa. This took less than an hour; a ground survey for the same purpose would have required several days.

Vehicle and foot surveys

It is usual to coordinate and combine surveys made from vehicles with those made on foot. One method is to drive variable distances, pausing at selected points to make either observations of adjacent rock outcrops or traverses on foot. The foot surveys may be several miles long, either returning to the same point, or picking up the vehicle at a different locality. Figs 119 and 121 show examples of maps of previously unmapped country resulting from vehicle and foot traverses planned initially from aerial photographs. The objects of these surveys were the location and development of borehole supplies for the towns of Dhala and Ataq both of which had relied on a few unreliable shallow wells.

Dhala (Fig. 120), situated near the South Yemen–North Yemen border, receives an annual rainfall of about 250 mm, but because this falls in one short season, the existing shallow wells were inadequate to meet the demands for drinking and irrigation. This region consists of a broad north–south complex of alluvial wadis at about 1400 m altitude, surrounded by steep, rocky crags and escarpments rising to 2500 m along the ridge of Jabal Jihaf. Fig. 121 shows the southern part of the Dhala valley.

Fig. 119 (opposite) A geological map of the southern part of the Dhala Valley, South Yemen. The valley is just over 1200 m above sea level; local craggy hills are 50 to 100 m high and the escarpments and the gorge have a relief of about 300 m. Boreholes were located to penetrate agglomerate (the aquifer) and to be up-dip of dykes.

The whole region is formed of late Cretaceous to early Tertiary volcanic rocks (the Aden Traps) which exceed 3000 m in thickness. They are mostly olivine basalt lavas with subsidiary interbeds of ash and agglomerate, but there are also trachyte lavas, pitchstones and ignimbrites. These rocks now dip southwards at angles of between 5°

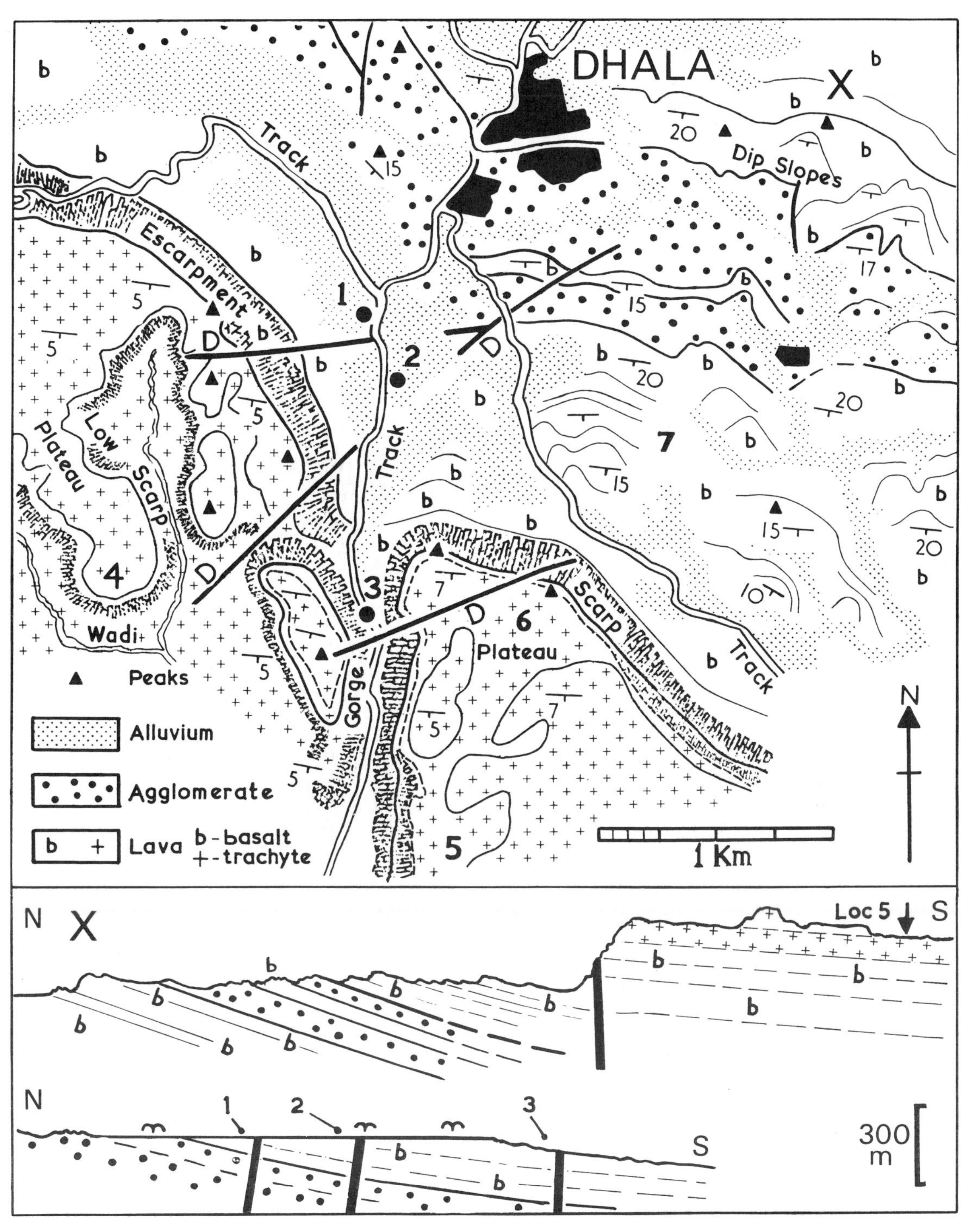
DHALA
X
Track
Dip Slopes
Escarpment
Low Scarp
Plateau
Wadi
Gorge
Scarp
Plateau
Track
Peaks
Alluvium
Agglomerate
Lava b-basalt +-trachyte
1 Km
N
N
X
Loc 5
S
N
S
300 m

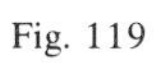

Fig. 119

A
B

and 20° and are cut by dolerite, trachyte and felsite dykes (Moseley 1969b). Recent erosion and deposition has resulted in the formation of wadis filled with alluvial sand and silt, usual thicknesses being about 10 m. Water wells in use at the time of the survey were entirely hand dug through the alluvium, with a 2 to 5-m sump excavated into the solid rock beneath. They therefore depended on recharge from the alluvium and the water table tended to fail during the dry season. At Dhala the main well for the town was found to contain only 1 m of water in which there were thousands of tadpoles and nearly as many frogs. It was not known if the solid rocks were aquifers, but it seemed at least possible that the almost uncemented agglomerates and ashes would prove so. Using the photogeological map as a basis several traverses were made to add lithological and structural detail to the map. It became clear that there would be annual recharge into the agglomerate, and that the relatively impervious dykes could act as underground dams. This meant that it was easy to predict borehole sites and depths to aquifers (from the outcrops and angles of dip). Borehole 1 (Fig. 119) predicted water at about 120 m, and a supply of 4000 gallons per hour (the maximum for the available pump) was obtained at this depth. This was a good supply for this part of the world and this well alone was able to give Dhala piped water pumped to a tank above the town (Shapland *et al*. 1967).

Ataq (Figs 121 to 123), situated in the northern part of South Yemen near to the Saudi Arabian border and to the extensive dunes of Ramlat Sabatayn, was a rather more difficult case which would have benefited from geophysical survey. The solid rocks are predominantly basement metamorphics with gabbroic intrusives, but in the east there are Mesozoic to Tertiary limestones and sandstones and in the north a prominent Quaternary volcano. Most of these rocks have been eroded into the north–south valleys, which were subsequently buried by alluvial fan deposits of sand and gravel, although the intervening ridges and other higher ground still stand clear. The basement rocks are largely impervious and water is contained in the alluvial deposits, which are recharged by annual rains of about 70 mm. Geological survey therefore required location of buried channels beneath the alluvium, where water would be concentrated. Seismic and resistivity survey could have defined these channels, but since neither was available channel positions had to be estimated from outcrops of solid formations further south (Moseley 1966). Suggested sites were just west of Ataq along an E–W line with a 200-m spacing, but South Yemen became independent before the borehole programme could be started and nothing more is known about it.

◁ *Fig. 120* Tertiary volcanic country in the mountainous region of Dhala near the border between South and North Yemen. **A**, A general view of the Dhala valley. The town of Dhala and adjacent villages stand on the craggy hills, with cultivation restricted to the alluvium-filled valleys. Military escorts were essential for all survey work. **B**, Gently dipping alternations of lava and tuff south of Dhala, with the lavas forming the escarpments of a typical 'trap topography'. The sharp peak to the left is a volcanic plug. In hydrogeological terms the loosely cemented pyroclastic rocks are the important aquifers. One misleading aspect in these arid regions is that most rocks are coated with black 'desert varnish', and during reconnaissance survey it is easy to misname lavas as basalt. Many of these lavas are in fact trachytic, black on the weathering surface, but much paler on freshly broken surfaces.

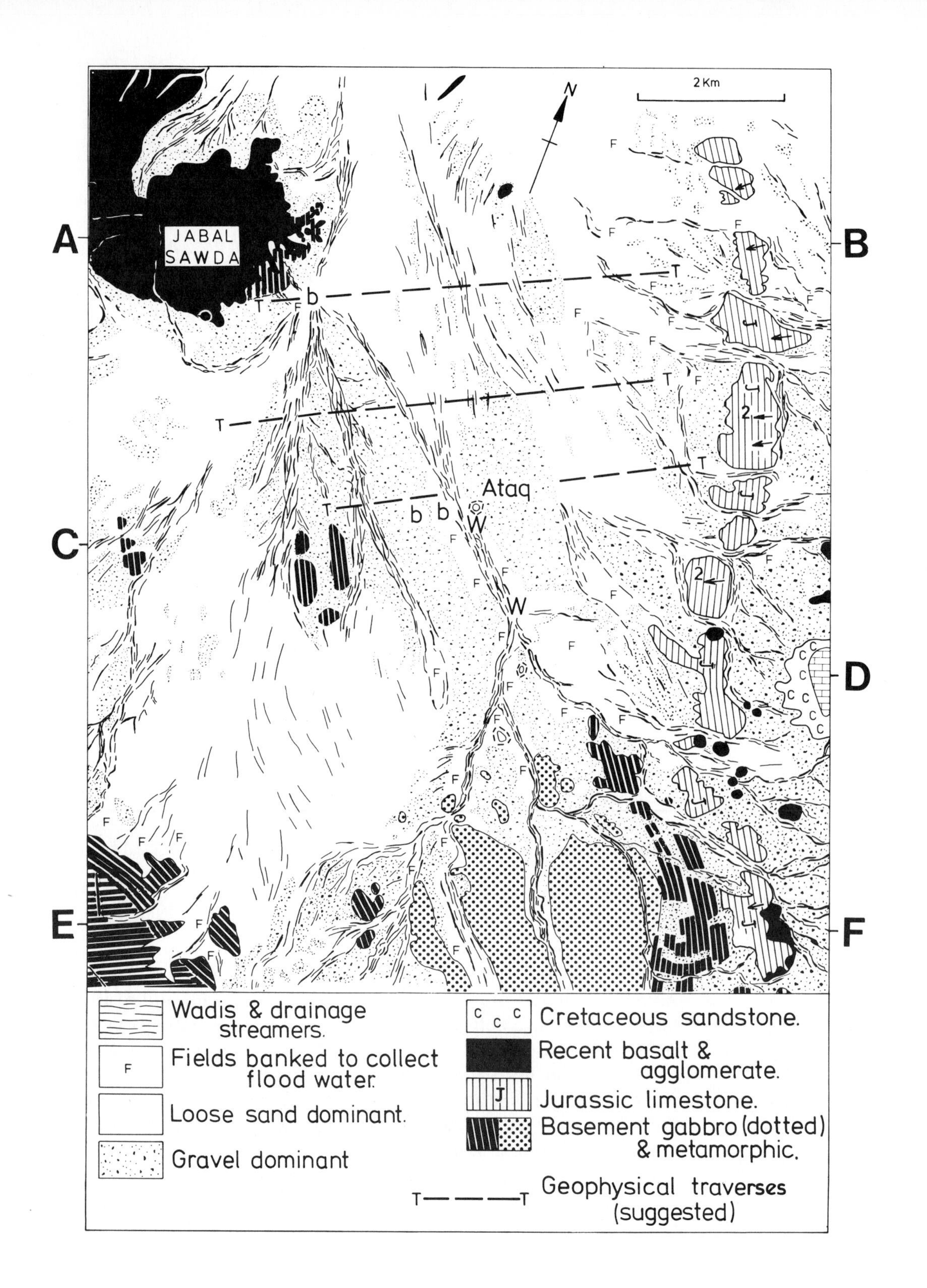

2 Km
N
A
B
C
D
E
F
JABAL SAWDA
Ataq
Wadis & drainage streamers.
Fields banked to collect flood water.
Loose sand dominant.
Gravel dominant
Cretaceous sandstone.
Recent basalt & agglomerate.
Jurassic limestone.
Basement gabbro (dotted) & metamorphic.
Geophysical traverses (suggested)

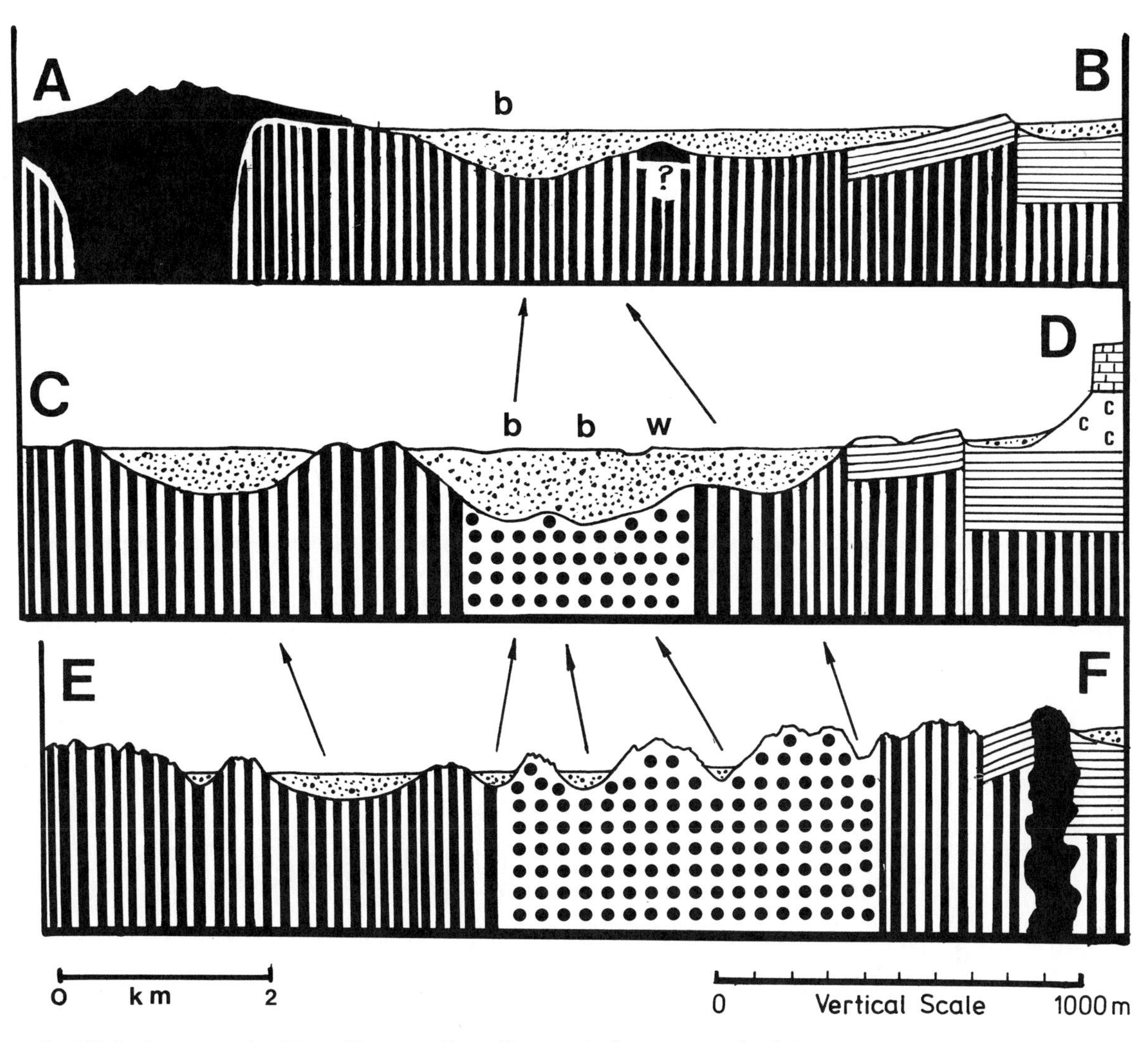

Fig. 122 Sections across Fig. 121. The locations of valleys and ridges are clear on section E–F, and their continuation can be predicted for the other two sections, although geophysical survey here would define the solid–alluvium boundary with greater precision. Ornament as for Fig. 121.

◁ *Fig. 121* Geological map of Ataq, South Yemen. Sections A to F are given in Fig. 122.

Preparation of reports for non-geologists

The object of the South Arabian surveys was primarily to locate borehole sites so that army well drillers could begin operations. It was also necessary to outline the geology in such a way that it could be understood by army personnel who had received a minimum of geological training. There is a certain skill in report writing of this kind, and it is a skill that most geology undergraduates do not possess, since they have spent several years learning to do just the opposite—that is to use a complex geological terminology. Nevertheless professional geologists frequently have to write reports along these lines; for example, an engineering geologist has to ensure he is understood by his superiors, who will probably be civil engineers whose university course in geology will have been short, elementary and largely forgotten. A sure guide in these cases is to avoid all jargon and to use only elementary names such as limestone, sandstone, lava flow, volcanic ash, etc., although even some of these may require a few words of explanation. For example 'granite' is not certain to be understood since it can mean any hard rock to the layman. Examples of undesirable and desirable phrases are:

1. *Not* 'thick deposits of alluvium', *but* 'thick alluvial deposits of sand and gravel' (the word alluvium is likely to be unknown but when it is followed by sand and gravel it becomes self-explanatory).

2. *Not* 'the sequence consists of basalt, trachyte, and related agglomerate and tuff', *but* 'the rocks consist of black basalt lava flows and related rocks, and also of rock fragments formed by volcanic explosion (agglomerate and tuff)'. The terms used in the first sentence are common enough for geologists, but they are jargon for non-geologists.

The second requirement for reports of this kind is for the writer to be quite definite about what is to be done. A well-drilling team will wish to know exactly where a borehole is to be drilled, how deep it should be taken, and the rock sequence they will penetrate. The following statement would not be welcome to well drillers, although it may be a true

Fig. 123 Oblique easterly view across the Ataq region (Fig. 121). In the foreground there are plains of Quaternary sand and gravel forming a broad alluvial spread (W) which forms the principle aquifer of this region. Beyond there are low hills of Jurassic limestone (J), and in the distance, scree-covered slopes of Cretaceous sandstone (C) surmounted by the steep escarpments of Lower Tertiary limestone (about 6 km away). Much further east in Wadi Hadramaut the sandstone is the important aquifer, and the limestone forms the extensive dissected plateau of the Jol.

representation of knowledge following a reconnaissance survey. 'There are distinct possibilities that boreholes south of Dhala would be successful, but the reconnaissance nature of the survey, together with extreme variability of the volcanic rocks in this region makes it impossible to predict depths to aquifers and yields with certainty.' It is far better to express the report as follows (assuming that there will be no legal action if it proves to be wrong). 'A borehole south of Dhala should be located at grid reference 707 133, 50 m SW of the junction of two tracks. It will pass through black compact basalt lava flows for 110 m (outcrops of the same rock can be seen 100 m west of this site), after which it will penetrate the aquifer, a porous fragmental volcanic rock (agglomerate) which will give a good yield. The standing water level will be at a depth of about 10 m.'

Finally there will be many occasions when reports have to be produced quickly, and this was the case in South Arabia as indicated above since the well-drilling team was on standby. The initial maps in such circumstances will be rough, with names, geological lines and ornament hand drawn on to tracing paper so that copies can be reproduced for distribution. This is an operation which newly qualified geologists should not find too difficult, since it is not unlike the writing of a tutorial essay or an examination answer.

18 Reconnaissance survey of the Lilloise igneous complex (Tertiary), East Greenland

Expeditions to the Arctic, Antarctic or to high mountains such as the Himalayas are likely to face similar problems to those described here, especially if funds are short. In these cases the scientists must expect to be labourers as well as geologists, and a large proportion of time will be occupied by these activities, and in travel to destinations.

East Greenland has been a happy hunting ground for geological expeditions since the early days of Wager and Deer, quite understandably since the geology is of fundamental interest, it is 'near enough' to the crowded geological environments of Europe and North America, and the exciting terrain caters for the most extravagant sense of adventure. The arctic conditions with high mountains, glaciers reaching the sea, and virtual absence of human occupation mean that expeditions have to be entirely self-sufficient. Organization has to be meticulous since parties may be isolated for many weeks, and of all the regions on Earth, only the Antarctic can offer more severe conditions.

The expedition referred to here was the 1971 Sheffield University expedition to Lilloise (Fig. 124), organized and led by Peter Brown. The primary intention was to examine the hitherto unvisited Lilloise Massif, suspected to be a major Tertiary intrusion. Although only 20 km inland this massif rises to 2500 m and is surrounded by heavily crevassed glaciers, so that it was anticipated that both approach routes and examination of the steep rock faces would present some difficulty. A degree of flexibility was necessary, however, and secondary objectives were to examine any rock outcrops which were accessible, since the geology was practically unknown. Of the six members of the expedition, four were from Sheffield and they (and particularly the leader) took on the whole of the organization so that all I had to do was to buy a few geological hammers, and catch a train to Newcastle. There is no doubt that the logistics before departure had meant a great deal of hard work for more than a year, and anyone wishing to arrange a similar expedition must expect to make initial enquiries perhaps two years before the intended date. It had been necessary to obtain a ship, and so the 130-ton wooden sailing ship *Signalhorn* had been hired to transport the party from Bergen to Greenland, the cost being shared by an American (University of Oregon) expedition of six led by Professor A. R. McBirney, who were going to Skaergaard (Fig. 124).

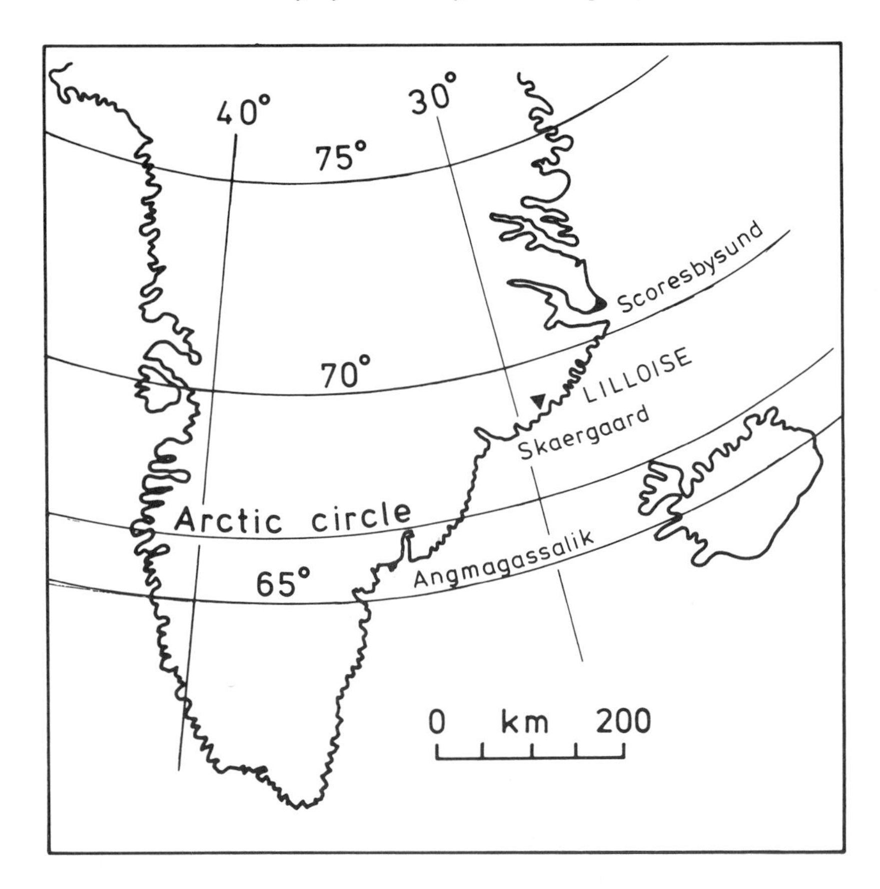

Fig. 124 Greenland and the location of the Lilloise Massif.

There were about 40 boxes containing stores and equipment; the latter included tents, stoves, sledges and all the climbing gear needed for travel in an arctic region of ice and rock. Once in Greenland all this material had to be transported up a glacier to a suitable camp, and since all this influenced the amount of geological work which could be done, if may be of interest to quote from some of the pages of a diary, which was kept in addition to field notes. This will be followed by a more detailed record of methods used during the fieldwork.

The expedition members left Tynemouth for Bergen on 14 July. The *Signalhorn* sailed from Bergen on 15 July, and five days later entered the pack ice north-east of Iceland. Part of the daily log of events is repeated in an abbreviated form below.

Tuesday 20 July. Enter the pack ice. Sea becomes very calm. Factory ships and trawlers nearby. Numerous birds and seals. Pack thickens and prevents progress. We have to back out and take a southerly course. Speed slow, about 2 knots. Generally low mist with the sun showing through.

Wednesday 21 July. We emerge from thick pack ice again and sail south-eastwards away from the Greenland coast. Try another lead, but it closes and the ship becomes immobile again. Mist over the pack clears at lunch time as the sun breaks through.

Thursday 22 July. Forced to run SW along the edge of the pack. Iceland now visible again about 20 miles SE. Bright sunshine, warm.

Friday 23 July. Another lead takes us within sight of the Greenland coast, perhaps 30 miles distant. Close pack halts the ship, but eventually slow progress is made towards Skaergaard. More open water near the Greenland coast. A whale follows the ship. Bright and sunny. Reach Skaergaard in the evening where the American expedition leaves the ship with equipment. Sail north immediately; water relatively clear near the coast.

Saturday 24 July. Pack ice north of Skaergaard thins out near to the coast, and progress is rapid. Locally it thickens where glaciers reach the sea, breaking off as large icebergs. Pack ice thickens again towards our destination at Wiedemanns Fjord. Becomes cloudy and cold.

Sunday 25 July. Wiedemanns Fjord. Low cloud, cold. Stores and gear rowed ashore in the lifeboat to establish a base camp on the beach (midday). *Signalhorn* leaves for Skaergaard where it will act as a base for the American party. In the afternoon some boxes of stores and the pulkas (one-man sledges) are back-packed across the moraine to a dump on the glacier. Clouding over.

Fig. 125 East Greenland. Sketch map of Lilloise and adjacent regions showing the main routes followed and the localities and peaks recorded in Figs 126 to 136.

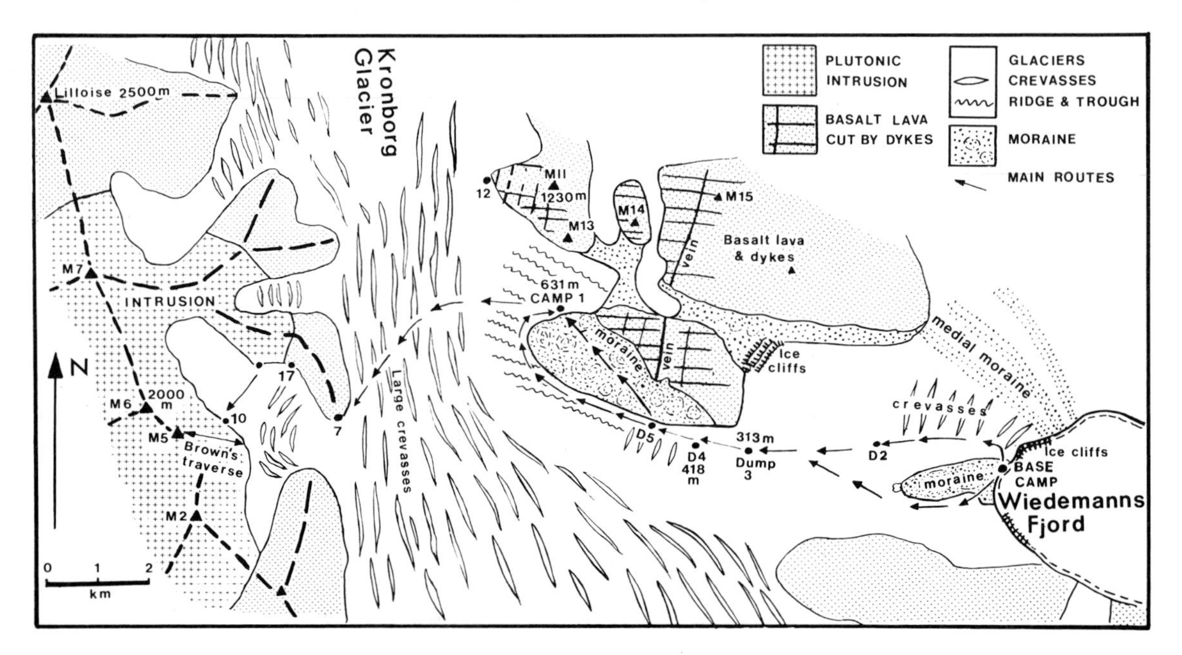

Monday 26 July. Transport of stores up the glacier, partly by back pack and partly by pulka (about twice the back-pack load can be moved on a pulka). Make a new dump at 'dump 4' on the glacier at 1400 feet [Fig. 125]. Return to base camp for another load. Partial cloud, walking up the glacier is warm work.

Tuesday 27 July. Carry on past dump 4 to establish a higher camp. Stores at dump 4 to be collected later. From dump 5 [D5 on Fig. 125] on the moraine at the edge of the glacier (1450 feet), stores, tents and personal kit are back-packed across moraine. This is formed of 100-foot high hummocks of large boulders, but all the hummocks are ice cored and unstable [Fig. 131]. The going is consequently arduous and the journey takes all day, but it is bright, sunny, warm and therefore not unpleasant.

Wednesday 28 July. Establish camp 1 at 2050 feet (aneroid). Sea level pressure rise today to 29.9. Locality 6 (field notes). Camp 1 is on the edge of the glacier west of the moraine. Sunny, but a cold (katabatic) wind blows down the glacier. It is necessary to make several journeys to dump 4 to bring up food boxes, the same boxes will be useful for transport of specimens on the way back. Some of these are brought by back pack across the moraine and some by pulka along the edge of the glacier. The latter is more efficient with twice the load carried, but there are difficulties as follows: (1) There are small crevasses near dump 5, some of which are awkward for pulka crossing. (2) Ice morphology of ridge and trough makes pulka pulling hard work in places. Ridges are about 4 feet high and it is almost 6 feet between ridge crests, easy work when ridges are parallel to the route but hard when they are transverse to it [see Fig 125].

Thursday 29 July. Cloudy with a cold wind, becoming sunny later. Reconnaissance walk across the Kronborg Glacier, which is about 5 km wide at this point. It is heavily crevassed, with crevasses 100 feet and more deep, but by zig-zagging it is possible to by-pass most of them, and others are wide, V-shaped, which makes it possible to climb into them and out the other side. The only difficult ones are those concealed by snow bridges, with the danger of collapse. However, the traverse is not difficult, the round trip, returning to camp 1 taking about 14 hours [Figs 126 to 129].

(The expedition diary was continued in similar fashion until the return to the United Kingdom.)

This diary reveals that 16 days elapsed from the time the expedition left the UK to the time it reached the first of the rocks to be examined. The six expedition members had in fact spent 14 non-productive man-weeks in travel and manhandling stores and equipment up the glacier, and at the end of the expedition a similar amount of time was necessary to transport specimens back to the ship and to travel back to the UK. Research grants for expeditions of this kind are generally from government sources and in this case the money available meant there was no alternative to the long sea journey and the manhauling on the glacier. The grants are pared down to a minimum, which is understandable considering the competing claims of other research work,

Fig. 126 East Greenland. A view from the slopes of M11 to Wiedemanns Fjord 11 km away. The positions of the base camp (B) and glacier dumps (D2 etc.) can be seen. The moraine in the foreground (M) is completely ice cored (see Fig. 131).

Fig. 127 East Greenland. A view from the slopes of M11 to Camp 1 (C1), the moraine (M) and Kronborg Glacier. The heavy crevasses can be seen in the central part of the glacier some 4 km from the viewpoint, with ridge and trough ice morphology in the foreground.

but surely this is a false economy. No matter how much members of such expeditions enjoy the experiences, the non-productive man-weeks constitute an expensive misuse of highly qualified scientific time. A much better use of research money in my opinion would have been to fly the expedition to a main base camp—this would probably have been possible had an extra £2000 been allocated (1971 prices). It would have been better still had a helicopter been available, but where there is no immediate economic, financial or political gain this may be asking too much.

Fig. 128 East Greenland. A view from the slopes of M11 to the Lilloise intrusion and the surrounding basalt. The route across the glacier was through the deep crevasses to locality 7, 6 km from the viewpoint, and thence to locality 10 (hidden). The Brown traverse was from locality 10 to M5, about 9 km from the viewpoint. The glacier M2 to M4 is a corrie glacier tributary to the main Kronborg Glacier (see Fig. 125).

Fig. 129 East Greenland. Crevasses and ice fall below M2 (part of the intrusion) illustrate some of the difficulties to be overcome before reaching solid rock (see Figs 125 and 128).

Survey methods

Once rock outcrops were reached, survey procedures were similar to the reconnaissance methods already described (chapters 9 and 16). Aerial photographs were used, but unfortunately the runs of vertical photographs were just to the north-west of the main survey area, and the expedition had to make do with obliques. These were supplemented by a 1:250 000 map (Danish Survey, 1931) which although small in scale, proved to be accurate and reliable. Once on the ground numerous stereophotographs were taken with a variety of lenses, both of distant mountainsides and of near rock outcrops. The photographs were accompanied by field sketches such as those of Figs 133 and 134, details being added after study of the rock faces using binoculars. These were particularly valuable when looking at distant mountainsides; for example, from camp 1, layering could be seen on mountains M1 to M6 8 km away (Fig. 128) which was invisible to the unaided eye. An aneroid barometer was also invaluable for height determination since, with such a short time available, the usual survey methods were out of the question. Normal precautions when using an aneroid are to have one instrument constantly monitored at a base camp so that diurnal pressure variations can be recorded, and the instrument readings made during survey are corrected on return to camp. This was not possible on our expedition since the whole party was engaged in the field for the whole time. In fact two members, Jack Soper and Rod Brown, went further up the glacier to examine the Syenite mass of Borgtinderne, with the remaining four forming a separate Lilloise party.

The most important results of the expedition came from the 1000 m vertical traverse from locality 10 to M5 by Peter Brown and Pat Fearnehough (Figs 125 and 130). This showed that the Lilloise intrusion, which was expected to be syenitic, in fact consists of layered cumulates ranging from peridotite at the lowest exposed levels to olivine–pyroxene–plagioclase cumulate at higher levels (Brown 1973). Other results came from observations of the basalt mountain M11, situated conveniently near camp 1, and the reconnaissance of Borgtinderne remotely situated at the head of the Kronborg Glacier. There were also purely incidental observations of glacier and moraine made during passage from one locality to another.

The method of recording the succession of M11 provides a good example of procedure (Figs 133 to 135). Initially field sketches were made and stereophotographs were taken from camp 1 (Fig. 133) and again from localities closer to the mountain (Fig. 134). Details of a few selected successions were finally recorded, of which locality 12 was one (Fig. 135). This showed that the basalt lava flows are often about 10 m thick with amygdaloidal tops and occasional red boles suggesting subaerial weathering, and that there are dykes of two trends. Other localities were examined in similar detail (for example, locality 26, Fig. 133). However, most of the localities shown on this sketch were examined in reconnaissance fashion only, sufficient nevertheless to collect specimens and take measurements, and to show that the pattern of lavas and dykes seen at locality 12 was the same for the whole mountain.

The glacier observations were particularly instructive in permitting comparisons with ancient glacial regions such as those of highland areas in Britain. Origins of morainic mounds, for example, can be seen to be more complex than might be supposed (Figs 131 and 132).

Fig. 130 East Greenland. Part of the Lilloise layered intrusion taken from near locality 7. The Brown traverse (Brown 1973) covered the continuously exposed section to the top of M5.

A
B

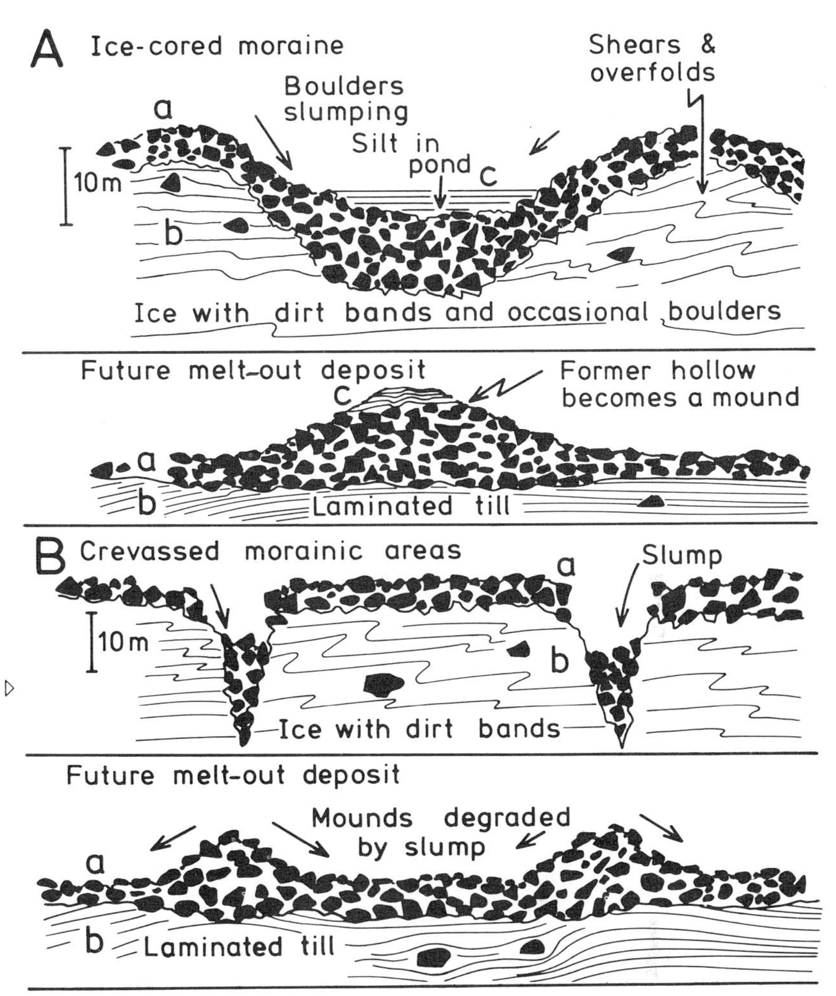

◁ *Fig. 131* East Greenland. Ice-cored moraine. **A**, Section about 10 m high showing a surface of large boulders underlain by debris-laden ice. In the upper part inclined shear planes can be seen, and lower down there are zigzag recumbent folds. These are ice-flow structures. **B**, A surface layer of large boulders underlain by debris-laden ice and an ice cave.

Fig. 132 East Greenland. Sections ▷ illustrating Kronborg morainic forms. **A** shows how hummocks of ice-cored moraine can become hollows following melt-out. Water-deposited laminated silt often seen on top of hummocks in British morainic areas can be accounted for in this way. The ice with dirt bands melts out into laminated till, easy to mistake for lacustrine deposits. In **B** crevasse infillings will result in irregular hummocky ridges. For further details see Boulton (1972).

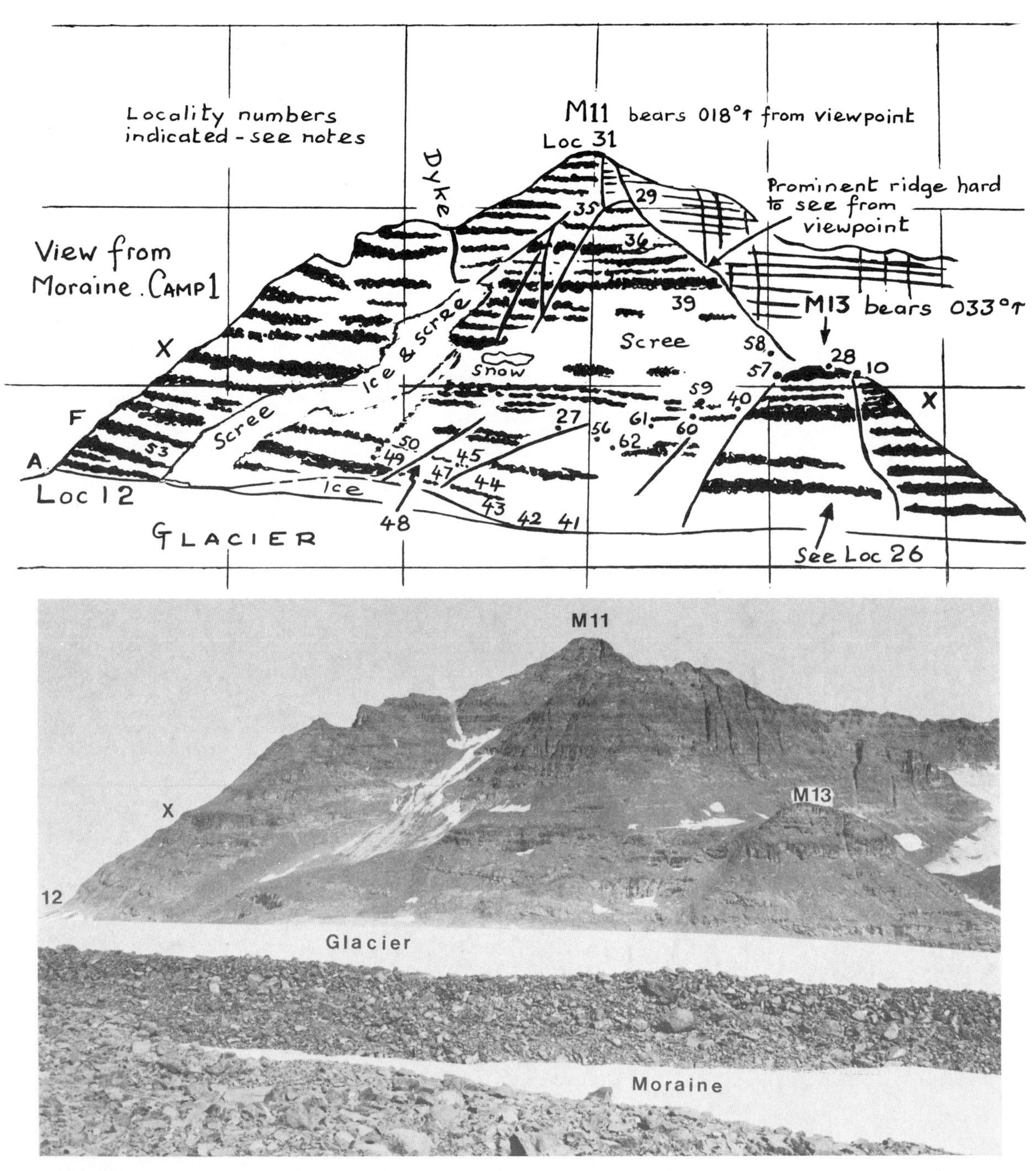

Fig. 133 East Greenland. Field sketch and one of the stereophotographs of mountain M11. Note that the field sketch is not completely accurate, for example the vertical relief is exaggerated—a common failing in the drawings of most people. However, it is sufficiently accurate for localities to be plotted, and for subsequent comparison with the stereophotographs, when the diagram can be corrected.

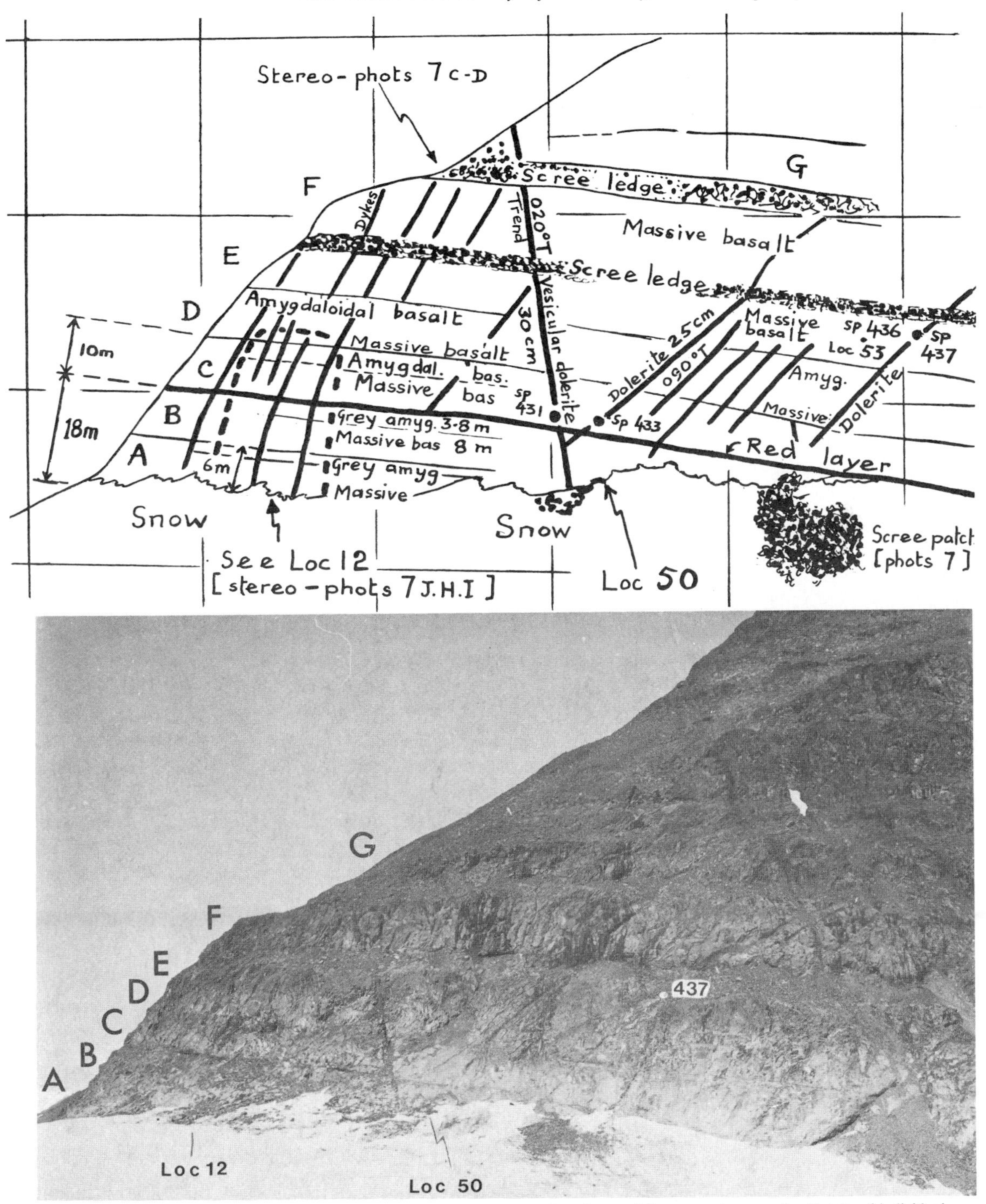

Fig. 134 East Greenland. A closer view of part of M11 with a field sketch and one of the stereophotographs. Details of individual lava flows and dykes can now be plotted, and although the field sketch contains inaccuracies, corrections can be made using the photographs. The latter would not be sufficient without the sketch.

Fig. 135 East Greenland. One of several stereophotographs of locality 12 (see Fig. 134). There were also field sketches of this locality and close-up colour transparencies of rock surfaces, and specimens were also collected. R, Rucksack; B and A, massive and amygdaloidal basalt; red layer, bole at the top of the lava flow; D, dolerite dykes; 1, 2 and 3, positions on the rock accurately measured from R.

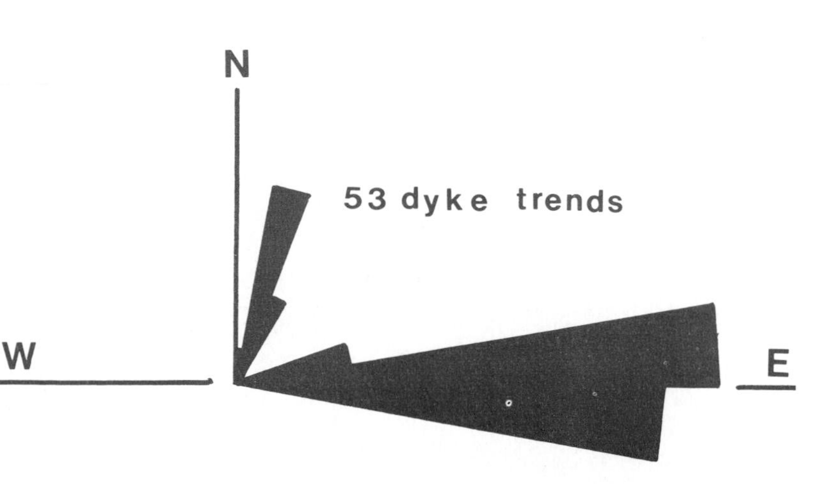

Fig. 136 East Greenland. Rose diagram of all dyke measurements from the area between base camp and M11.

19 Extreme reconnaissance of routes traversed in East Africa and Libya

It is often the case that a geologist will travel quickly across country adjacent to, or related to that which is being mapped, perhaps just to pass from one camp to another, or to return to a main base camp. Such journeys may be by air, by road or on foot, and the distances may range from a few to hundreds of miles, and may be in any part of the world, with climatic conditions varying from extreme cold to extreme heat. Usually such journeys are wasted because no permanent record of observation results. This clearly need not be so since the traveller can easily keep a systematic log of observations and localities which may subsequently be extremely useful. Examples of vehicle traverses are given below.

Land Rover journey, Kenya

This particular journey was from Nairobi to Magadi, in the Gregory Rift and was made on 4 August 1961. The journey was part of a terrain study which had already covered large areas in Kenya, Uganda and Tanzania, and would be applicable, for example, to highway construction. Consequently information was required not only on the geology and geomorphology, but also on soils and vegetation. Travelling as a passenger it was possible to make observations from the vehicle, and to record the vehicle mileage for each observation. Admittedly many such observations are likely to be in error because they are made quickly in a jolting vehicle, but they do provide a reasonable record of a journey which would subsequently have been largely forgotten. Occasional short stops were made at particularly interesting localities and photographs were taken both at the stops and from the vehicle. The notes for part of the journey were as follows:

Nairobi 9 End of tarmac. Dry grass. Scattered acacia. Ngong Hills ahead. Rather flat with valleys incised about 50 feet.

Nairobi 13 Gently undulating. Grass. Scattered bush. *Leonotis labiata* common (specimens of vegetation had been collected during survey further north and identified at Nairobi Arboretum, making identifications listed herein relatively easy).

Nairobi 21	West of Ngong Hills.
Phot. 204	View SW across (fault?) scarp.
Phot. 205	*Dombeye rotundiflora*; *Rhus natalensis.*
Phot. 206 / Specimen 206	Lava (? porphyritic aegirine basalt, see memoir). Spheroidal weathering. Weathered at outcrop. Massive unweathered below at base of flow.
Nairobi 24	
Phot. 207	Zebra. (Also giraffe.)
Nairobi 26 Phot. 208	Descend scarp (cf. Derbyshire 'edges'). Probably one of the rift faults. Solid rock (basalt) well exposed, but most slopes not excessively steep. Road rough.
Nairobi 28 Phot. 209	Near chestnut soil. Black surface (A) horizon (specimen 209A). Carbonate enrichment increases from 9 inches to a maximum 2 feet 6 inches below surface (specimen 209B). Rests on basalt.
Phot. 210 / Specimen 210	Basalt: feldspar weathers to calcite. Calcite coats joints.
Nairobi 30	Masai with cattle. *Tarchonanthus camphoratus* common.
Nairobi 32 Phot. 211	Boulder strewn surface. Basalt boulders mostly 1 to 2 feet in diameter cover a flat surface. Boulders form an almost continuous cover.
Nairobi 36 Phot. 212	Fault scarps. Steep boulder scarps about 50 feet high (basalt) with flat stretches between escarpments. Note that references to basalt are to hand specimens only.
Nairobi 39	Road corrugated. Speed preferably more than 40 m.p.h. to ride across the corrugations.
Nairobi 42	
Phot. 213	Road improves.
Phot. 214	Flat. Legamunge lake deposits (see memoirs and maps which record them as diatomite). Volcano (Olorgosailie) to the south.

Observations as listed above were continued in the same way for another 50 km to Magadi.

Land Rover reconnaissance, Libya

In this case the survey was a rather specialized and provisional hydrogeological reconnaissance (Moseley and Cruse 1969) and was concerned with virtually uncharted and rainless desert of the Libyan Plateau, the northern part of which was well known to the desert armies during the Second World War.

The area shown on Fig. 137 is extremely flat, but there are occasional low escarpments and shallow mudpans. It is stony desert (serir) with a thin cover of gravel, underlain by horizontal chalky to shelly Miocene limestone. The intention of the survey was to examine existing borehole sites and to suggest new ones. Ground water in this region relates to the 'main' or 'sea level' water table which is almost always present, falling from near sea level at the coast, to about 50 m below sea level 300 km to the south (Shotton 1946). This generally places it about 200 m below the level of the Libyan Plateau. Unfortunately this water is usually highly saline and our problem was to locate slightly less saline spots. Theoretically these are to be found where there are combinations of prominent mudpans and faults (generally expressed as straight escarpments), since occasional surface floodwater will be concentrated at the mudpans, which are essentially inland drainage areas with wadis leading to them, and underground flow will be facilitated by the faults (there are no salt pans since the water sinks into the limestone before it can evaporate) (Fig. 138).

The logistics of this survey (undertaken in April 1968) may be of some interest, mostly as an indication of what not to do in the desert. Two ancient Land Rovers and trailers were available and were loaded with petrol, water and food. There are no roads but the flat terrain is eminently suitable to Land Rover travel (Fig. 139), and even escarpments presented no problem. However, after 100 miles one trailer snapped into several pieces after being towed up an escarpment of chair-sized boulders. The wheels were cannibalized just in case. There followed a succession of punctures with the trailer wheels coming in most useful since none of the party seemed capable of mending punctures successfully. Nor were any of the party good mechanics so that it was just as well there was no engine failure. It says much for providence that this leg of the survey, which lasted about 10 days, was successful. Readers will be aware that to break down in the desert can be dangerous as well as annoying.

Since an area 300 by 200 km had to be covered in a short time our method was essentially the same as that just described for Kenya, but with one difference. The vehicles were driven in straight lines along compass bearings from one known point, usually from one large mudpan, to another. The use of a magnetic compass in this way involves a very necessary correction for vehicle error. The compass has to be 'swung' in the same way that an aircraft compass is swung and the error for every 30° of heading is recorded in graphical form so that the vehicle can be driven along the correct bearing. Alternatively a sun compass is quite suitable for general navigation in the desert. The survey legs varied between 15 and 100 km in length, with observations tied to vehicle mileage. This was an exercise rather like the navigation of a ship at sea, and in fact similar astronomical methods of navigation could have been used satisfactorily had there been more time to prepare for the survey. Location of position in this way is quite simple,

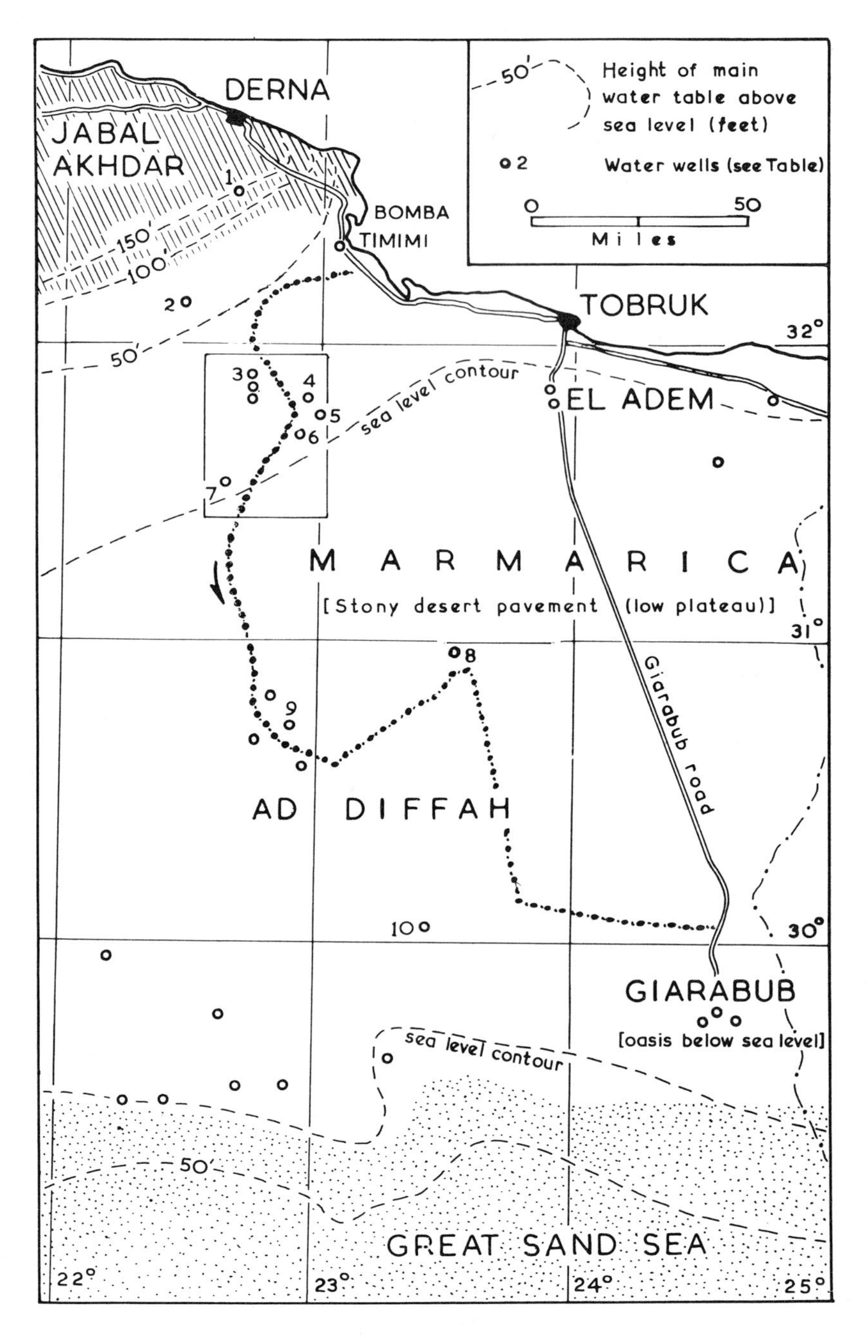

Fig. 137 North-eastern Libya showing the area of stony desert between Jabal Akhdar and the Great Sand Sea. The general route of traverse is shown as a dotted line and the rectangle shows the area of Fig. 138. (After Moseley and Cruse 1969.)

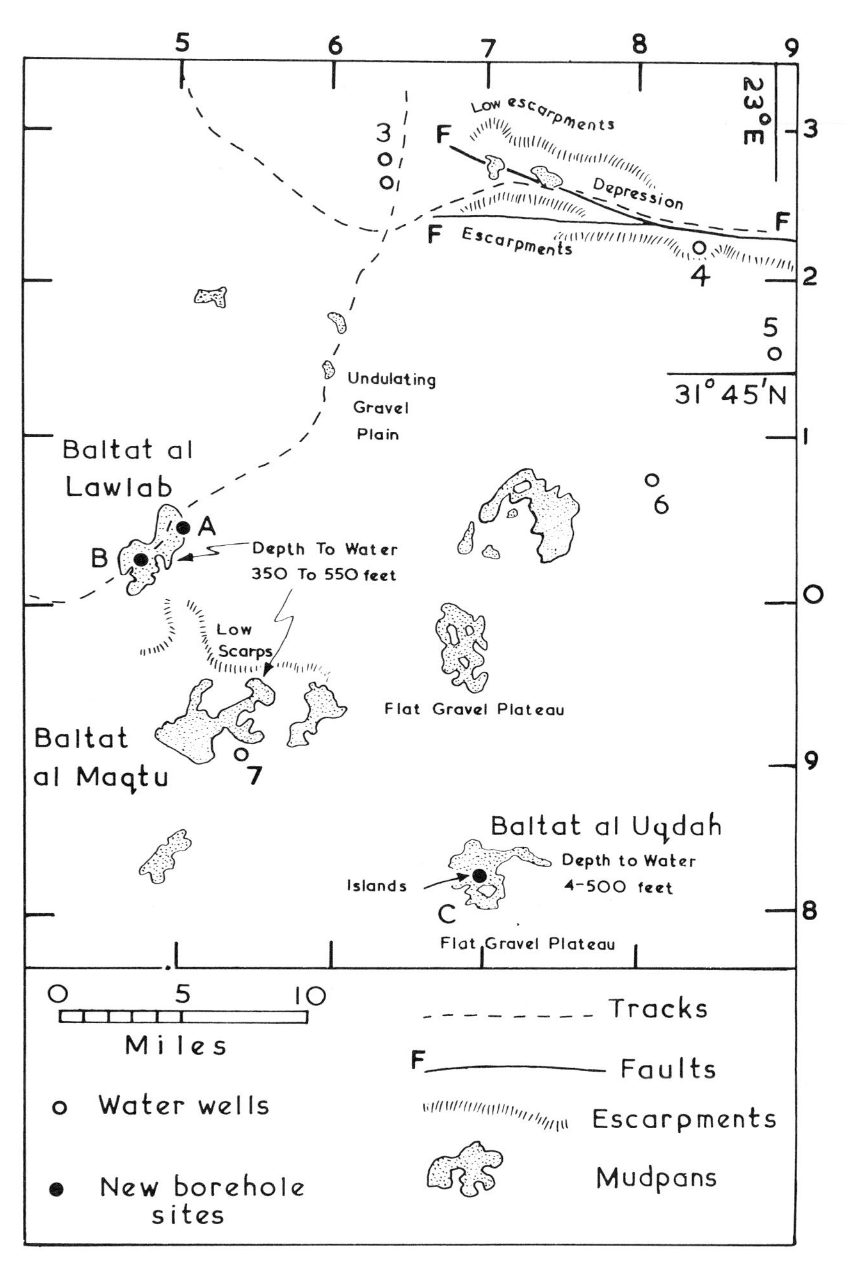

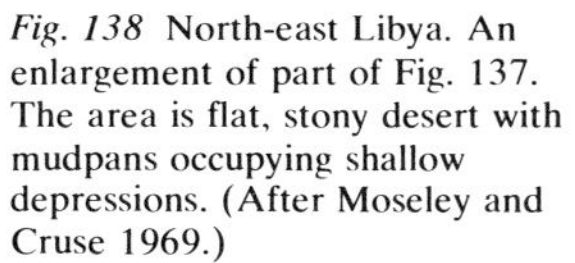

Fig. 138 North-east Libya. An enlargement of part of Fig. 137. The area is flat, stony desert with mudpans occupying shallow depressions. (After Moseley and Cruse 1969.)

but does depend on a certain amount of equipment and instruction. A sextant, accurate watch and a suitable radio to obtain time signals are necessary. A number of heavenly bodies are then observed from different quarters, each one giving a position on a small circle, with intersections locating position. In fact declination gives the latitude, and the hour angle from Greenwich Meridian gives the longitude. Empirical tables can be consulted, making the whole operation simple. It requires a few days' training only before an observer is able to locate position to within 400 m, and this is considerably more accurate than would be possible by following a compass bearing across 100 km of trackless desert.

More recently location methods have become available using orbiting satellites. They appear to be extremely accurate, and are likely to become standard equipment for reconnaissance surveys. In fact the methods become more refined year by year and can now be used for detailed land survey. There are several satellites in near-circular polar orbits at between 850 and 1100 km that are moving transmitters in space, and are in contact with a large area of the Earth's surface at any one time. Recording of successive satellite positions permits determination of latitude, longitude and height. The field receiver is a microprocessor system in a robust box about 50 × 40 × 20 cm, easily carried in a vehicle, which can also act as its power source (12 volt battery). Its operation is automatic once set up and it is provided with a geodetic positioning programme in Fortran IV language. Multiple satellite passes are taken and the results transformed on site, into latitude, longitude and height. Our small expedition of 1968 was not, of course, provided with such a luxury.

Fig. 139 Characteristic scenery of the north Libyan desert. **A**, Surface of calcrete forming an extensive limestone pavement (see also Fig. 96). Superficial subsoil formations of this type are easily mistaken for solid rock, and they can conceal important structures and lithologies in the solid rock. **B**, Survey vehicles can drive in any direction across extensive plains of this kind. In the foreground a gravel skim rests on calcrete, and beyond are gentle linear depressions which receive run-off from the occasional rains, and in consequence have a little more vegetation.

20 Geological survey of metamorphic terrains

Fieldwork in areas of metamorphic rocks requires a different technique from that used with most other rock types, largely because the rocks are structurally complex. Indeed much of the field observation will be concerned with structure, the variety of which is so great that it cannot possibly be dealt with adequately in this short chapter. There are, however, several excellent textbooks (e.g. Hobbs *et al.* 1976; Ramsey 1967) which students contemplating geological survey of metamorphic rocks should read. It is also necessary to be able to identify common metamorphic rocks in hand specimen, and to know something of their origins; after two years or so at university students should have acquired much of this knowledge. In spite of the difficulties inherent in this type of survey, the fundamental methods of observation that have already been described still apply. Structural measurements must be recorded systematically (see Figs 10, 22, 23, 24, 30, 31, 34, 38, 60, 91, 92, 95), but field identifications of rocks should not be relied upon too heavily. Numerous accurately located specimens should be collected for subsequent laboratory examination (see chapter 12). Field procedure will vary according to the part of the world being investigated, and to the nature of the survey (for example, a reconnaissance or a detailed survey), but generally there will be subdivision into initial photogeology followed by field survey, the latter consisting of the determination of lithology, the plotting of boundaries and the recording of structural details.

Photogeology

Most metamorphic rocks are steeply inclined and are comparatively resistant to erosion, so often form high ground. Structures are generally easily seen on aerial photographs as they form strong lineaments which include foliation, schistosity, gneissic layering and numerous fractures (faults and master joints). There is no problem therefore in plotting major structures from aerial photographs (Figs 35 and 140), but minor structures will not be so obvious—indeed they are unlikely to be visible—and it is frequently these structures that provide the key to structural interpretation. In this connection too much reliance should not be placed on aerial photographs. The texture and tone of the photographs also help in the gross differentiation of lithologies. Hard or competent rocks, such as quartzite or marble, can usually be distinguished from schists, which provide far more lineaments parallel to schistosity; and gneisses with more regular and coarser layering (Fig. 35) are easy to distinguish from schists. Fig. 140 also shows how synorogenic granite layers can be differentiated from gneisses and the boundaries can then be plotted. The latter rocks exhibit strong, parallel and close spaced lineaments (the gneissic layering), whereas the former are massive with a completely different texture. The granite can also be seen to transgress the gneissic layering in places. The importance of tone is evident on Fig. 13 on which acid and mafic rocks are easily differentiated by colour. All these photographic variations are no more than reconnaissance subdivisions, however, and on Figs 35 and 140 it is not possible to say whether the narrow darker bands are biotite gneisses, amphibolites, pyroxene granulites or some other dark coloured variety of metamorphic rock; nor can it be said whether the pale bands are acid gneiss, migmatitic granite layers or quartzite

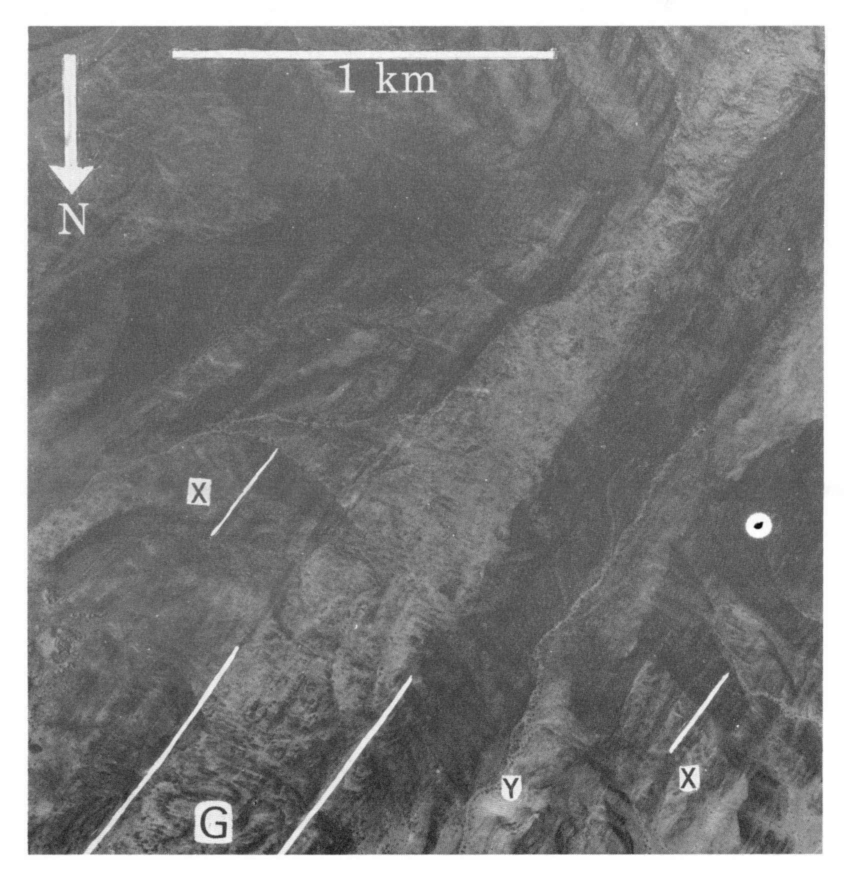

Fig. 140 Aerial photograph of basement metamorphic rocks near Beihan, South Yemen (see Fig. 35). The region is predominantly gneiss (X, Y), with a thick sheet of synorogenic granite(G). The gneissic layering contrasts with the photograph texture of the more massive, unlayered granite. Variations in layer orientation can be seen near Y. (See Moseley 1971b.) (Crown copyright/RAF photograph.)

interbeds. Determination of these details must be left to the ground survey. Nevertheless there can be no doubt that photogeology can be of the utmost importance as a preliminary to survey of metamorphic rocks.

Lithological boundaries plotted in the field

The photogeology has to be followed up in the field whether reconnaissance or detailed mapping is intended, and one of the more important tasks will be to plot on to a map the boundaries between different rock units, which may or may not be a confirmation of the photogeology. The principles are no different from those discussed in other parts of this book; the ground survey needs to determine not only the positions of the outcrops of these boundaries, but also their nature— whether abrupt, transitional or unconformable, or whether they represent intrusive or tectonic contacts. Reconnaissance surveys may require only one traverse across the strike, with specimens collected wherever necessary (Fig. 35); a detailed survey requires much more than this, and most boundaries have to be followed and studied carefully. For example, the contact of the major granite sheet on Fig. 140 would require careful study, especially where it appears to be transgressive. Likewise the nature of the alternating pale and dark narrow bands and their junctions with each other would require careful scrutiny. In this case there are hornblende and biotite gneisses alternating with acid gneisses and thin synorogenic granite sheets, of similar composition to the thick granite sheet just mentioned (Moseley 1971b). These matters cannot be ascertained entirely from aerial photographs and have to be decided by ground survey. It is invariably necessary to collect carefully localized specimens since even a metamorphic specialist cannot be sure to identify many of the minerals in the field. Ground photographs are equally necessary; many should be 'close-ups' since it may require a 1-m diameter specimen to reveal the true nature of the gneissic layering, and there is a limit to the number of these which can be carried in a rucksack.

Structural observations

Methods of recording structure have been described elsewhere in this book (Figs 10, 12, 22–4, 30, 31 and 34) and the same principles apply, even with high-grade metamorphic rocks. An objective in the field should be to make a quantitative record of different types of structure so that they can be plotted and analysed (usually stereographically). The crucial difference between structures in metamorphic and sedimentary rocks is that the former are almost always more complex and of greater variety. It is therefore of the utmost importance that students should not embark upon their own field survey before: (1) comprehensive reading of an up-to-date textbook, which should be a 'field companion' (e.g. Hobbs *et al.* 1976; Ramsey 1967); (2) completion of a field course dealing with the structural mapping of metamorphic rocks. Most university geological departments hold courses of this type; one week would be a minimum requirement (Figs 140 to 145).

A major difficulty on first acquaintance with metamorphic structures is to know what should be observed and measured. It is one thing to read about lineations in a textbook and another to realize the significance of some of the apparently obscure markings on the rocks when they are first seen in the field.

Planar structures

Planar structures are diverse, and may include original bedding, a variety of different cleavages, schistosity and gneissic layering, as well as numerous fractures (faults, joints, veins, etc.).

1. Original bedding can often be seen, more particularly in low-grade rocks. It should be measured as indicated on Fig. 34. In high-grade rocks the bedding is likely to become obscure or completely obliterated.

2. Cleavages of slates and phyllites may be of constant orientation presenting no measurement problems, but if there has been polyphase deformation there may be more than one cleavage, and the earlier cleavage may have been folded. There may also be several cleavage varieties such as fracture cleavage, slaty cleavage or crenulation cleavage that will reflect the style and history of deformation.

3. Schistosity and gneissic banding, like the cleavage of slates, is likely to be more complex than first acquaintance would suggest. Polyphase deformation is in fact the rule, but it is not easily detected by the inexperienced geologist.

4. Axial planes of folds should be measured wherever possible since they usually give more important information than (say) the measurement of bedding planes. They are not always easy to measure, however, and are far less numerous than other planar structures.

Folds

It is necessary to be able to recognize the different types of fold commonly seen in metamorphic rocks, and to be aware of the interference patterns resulting from polyphase deformation (Ramsey 1967). It is not possible to go into details here and in fact this is not necessary since all are described fully in standard texts and papers. A list of common fold types is as follows. (1) concentric and similar folds; (2) conjugate folds and the related chevron folds and kink bands; (3)

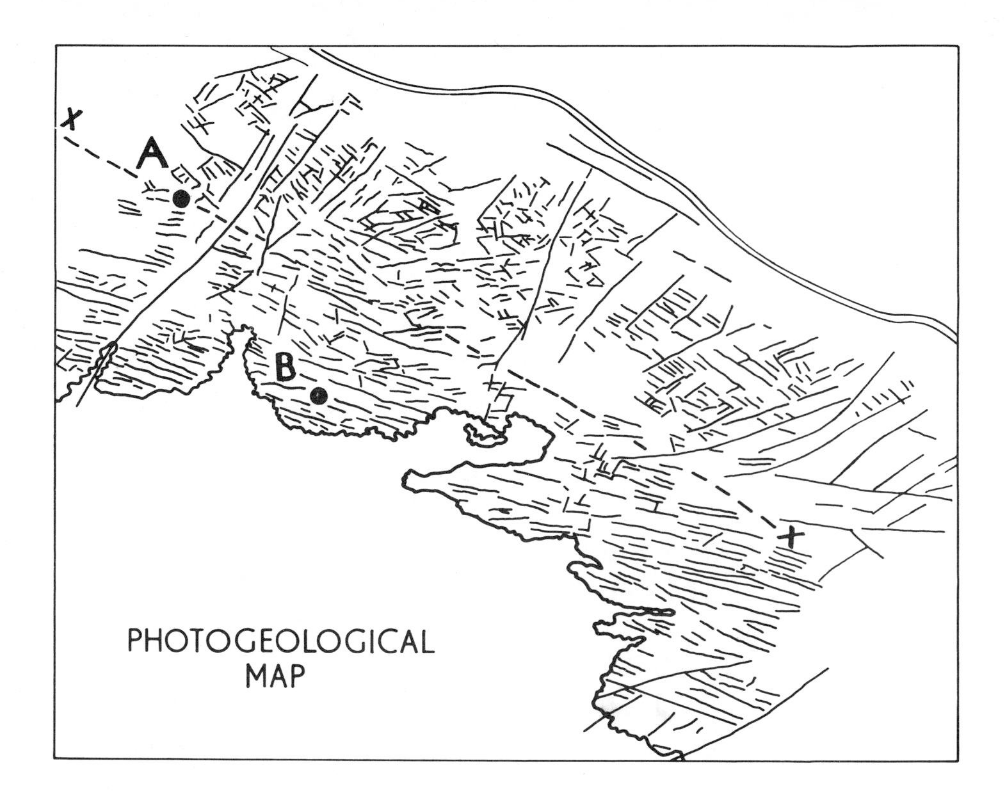

X
A
B
X
PHOTOGEOLOGICAL
MAP

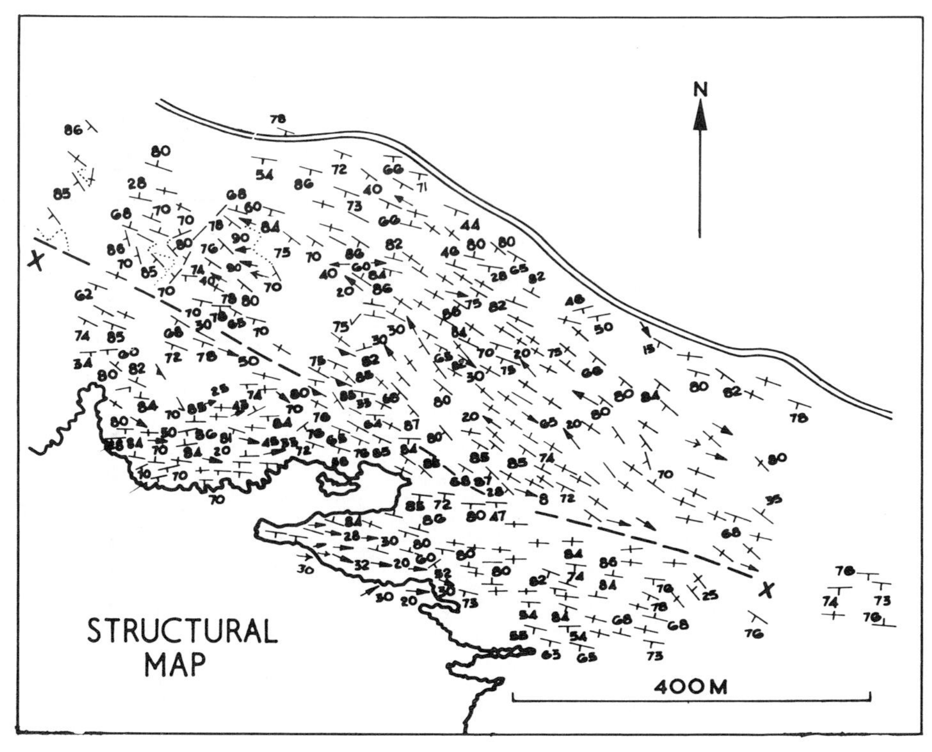

N
X
X
STRUCTURAL
MAP
400 M

◁ *Fig. 141* Photogeological and structural maps for Lewisian Gneiss, near Achmelvich in NW Scotland (Sheraton *et al.* 1973). The information was plotted by Birmingham University undergraduates during a structural mapping course. It will be noticed that the photogeological map suggests convergence of the gneissic layering to the north and south of X–X, and this is confirmed by the stereographic projections (Fig. 143). The scale of this map is too small for the details that are required in some particularly interesting areas, and tape and compass enlargements were necessary (Fig. 142). (See Fig. 142) for the legend.)

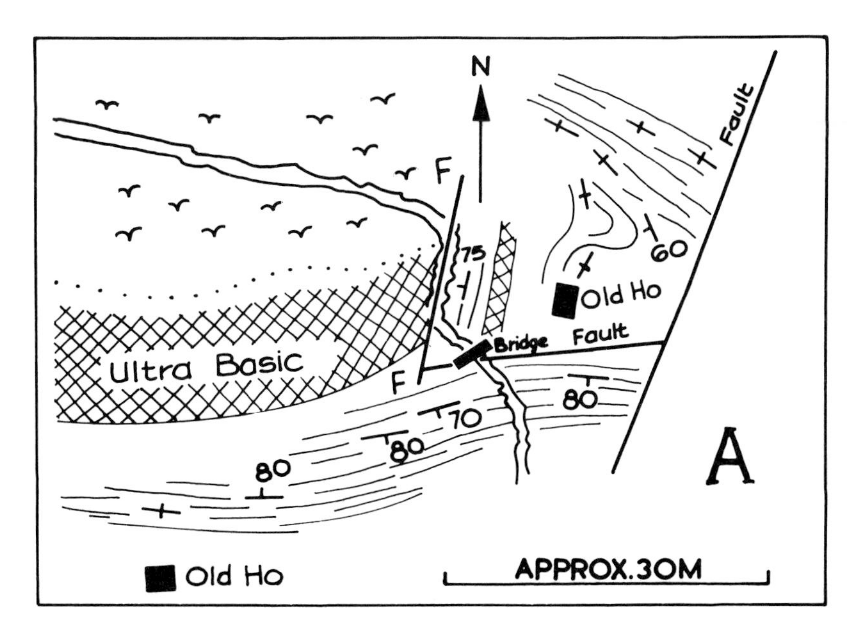

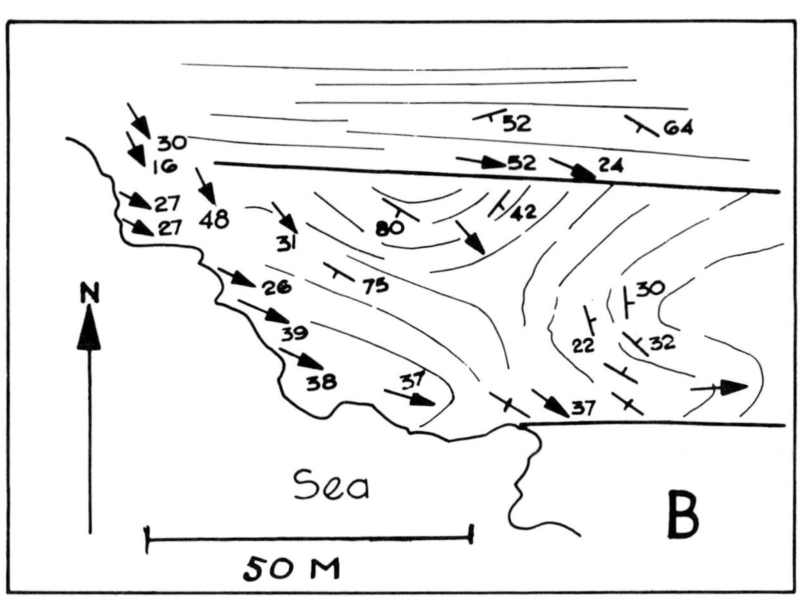

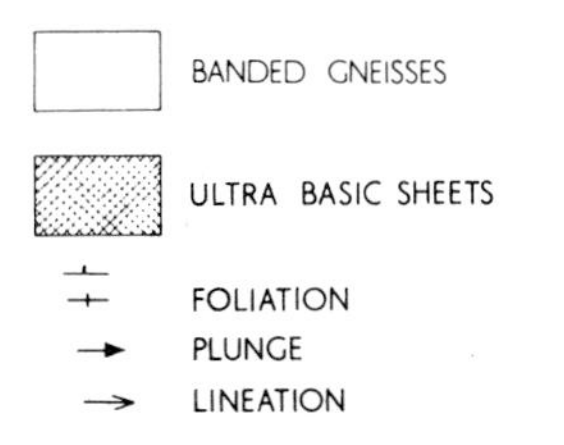

Fig. 142 Enlargements of some of the more interesting areas of Fig. 141.

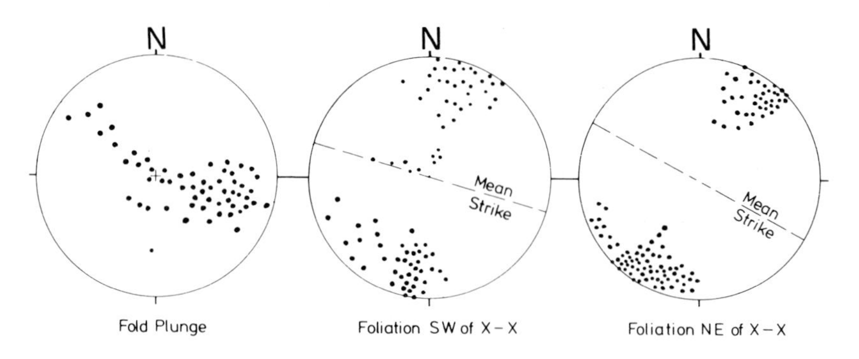

Fig. 143 Equal-area stereographic plots of structures from Fig. 141. It will be noticed that the patterns for the northern and southern areas are different, illustrating that initially it is important not to plot all observations for a large area on to one diagram.

disharmonic folds; (4) drag folds; and (5) many structures prove to be conical when their three-dimensional form can be seen. Polyphase structures are commonplace but in the field it is less common to locate refolded folds, such as those of Figs 144 and 145, than to deduce this by measuring numerous fold styles and plunges. The latter may be inclined in various directions (Fig. 146). It follows that the geometry and structural history of folded rocks of this kind are determined by taking numerous measurements, supported by field sketches and photographs. These measurements should include orientations of both limbs, the axial planes and plunges (see Figs 34, 95 and 100).

Lineations

Parallel orientation of linear structures is commonly seen on planar surfaces of metamorphic rocks. They may be penetrative (hornblende crystals in schist) or non-penetrative (most slickensides). The more important of these structures are listed here but, as with other structures, more comprehensive descriptions will have to be obtained from the specialist structural textbooks already referred to.

Slickensiding. This is extremely common as striations on planar surfaces and can give an indication of relative movement. The plunge (or the pitch) of these structures should therefore be measured, but it is important not to draw far-reaching conclusions from these orientations since a slickensided surface can be produced by a final surge of movement and may not be completely representative of the structure. For example, vertical slickensiding can be found on faults known from other evidence to be (strike-slip) wrench faults.

Intersection lineations. These are formed by intersections of planes of different orientation. In many slates cleavage is axial–planar to folds and the intersection of bedding and cleavage results in lineations parallel to fold axes. It is important to measure plunges of these lineations, either on bedding planes (where the cleavage forms the lineation) or vice-versa on cleavage planes. Caution is necessary since cleavage is not always parallel to the axial planes of folds (Fig. 10). Other intersection lineations are also common, such as those between two cleavages, which may result in pencil slates.

Fold hinges. Axes of folds are important linear elements and occur on all scales from regional to microscopic. Crenulations, especially well seen in hand specimens of schist, are important field structures. Often there will be intersections of crenulations of different orientation, which will yield information on the polyphase history of an area.

Mineral lineation. Naturally elongate minerals such as hornblende are often found with strong preferred orientation. This shows clearly on rock surfaces. Frequently mineral lineation is parallel to fold axes, but this is not always the case, and care is necessary before particular meanings are attached to orientations.

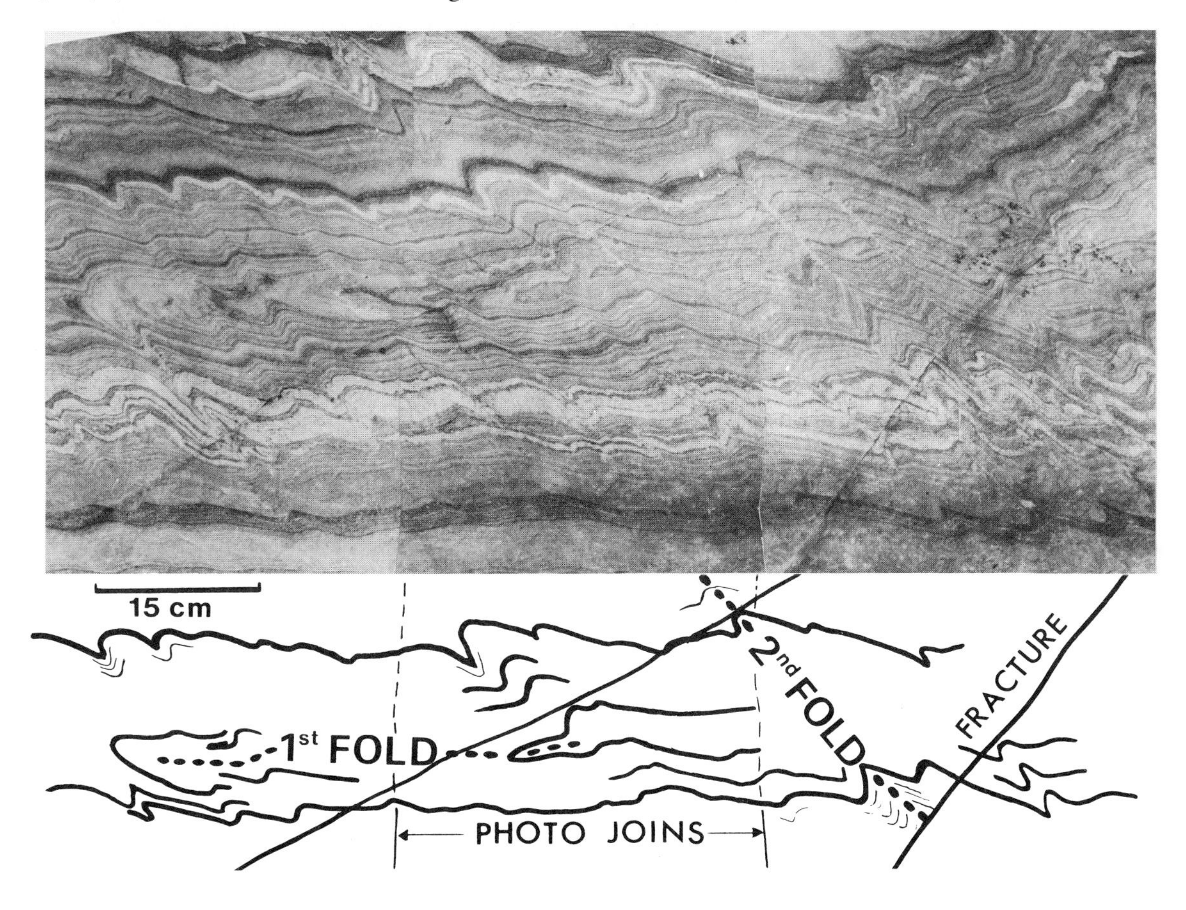

Fig. 144 Interference patterns of superposed folds, Loch Monar, NW Scotland (Ramsay 1958). This is a horizontal outcrop that is always difficult to photograph because, without a balloon or some other airborne conveyance, one is too near to the outcrop. The difficulty can be overcome by covering the area with overlapping vertical photographs to form a mosaic (compare with an aerial photograph mosaic). In this figure, taken on a Birmingham University field excursion, there are only three photographs but in other cases there may be up to 100 photographs.

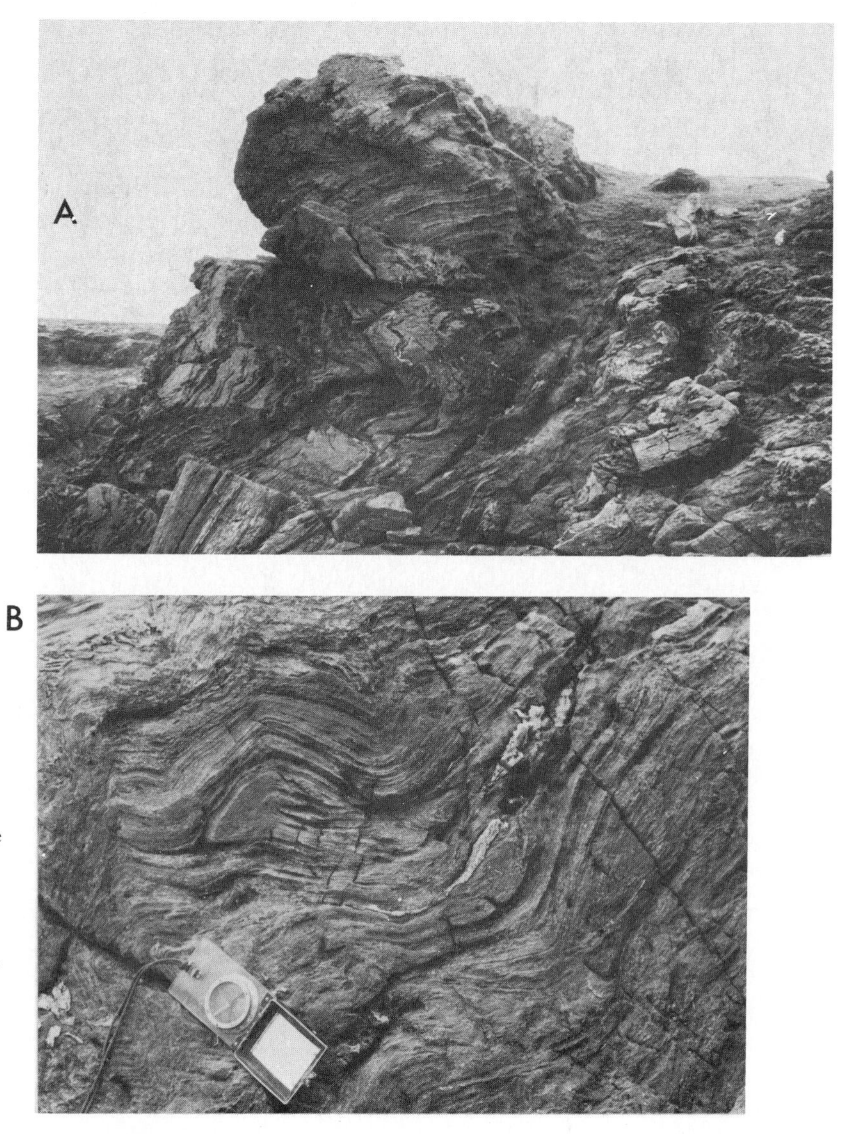

Fig. 145 Outcrops of glaucophane–epidote schist on the Isle de Groix, Brittany, France examined during a Birmingham University structural mapping course. **A** shows complex folding in the sea cliffs, but details of structure cannot be seen. **B** shows some of the detail in the form of superposed folds. Outcrops have to be examined carefully and quantitatively, since although structures such as those of **B** may be suspected, relatively few are unequivocal.

Fig. 146 Copy of a page of field notes by Don Aldiss on the Skiddaw Slates of Cumbria, northern England (Birmingham University BSc thesis, 1974). There are three fold phases represented in these outcrops, but no interference patterns are seen, and these polyphase events can only be confirmed by reference to other exposures in the region. The steep plunge of the folds presents the dip of a fold limb during phase one. Phase two is shown by the tight ENE folds and related cleavage, and phase three, open recumbent folds, can only be seen on vertical surfaces, where the other two phases cannot be seen. Similar surveys covering a wider area are recorded by Jeans (1974) and Webb (1972). ▷

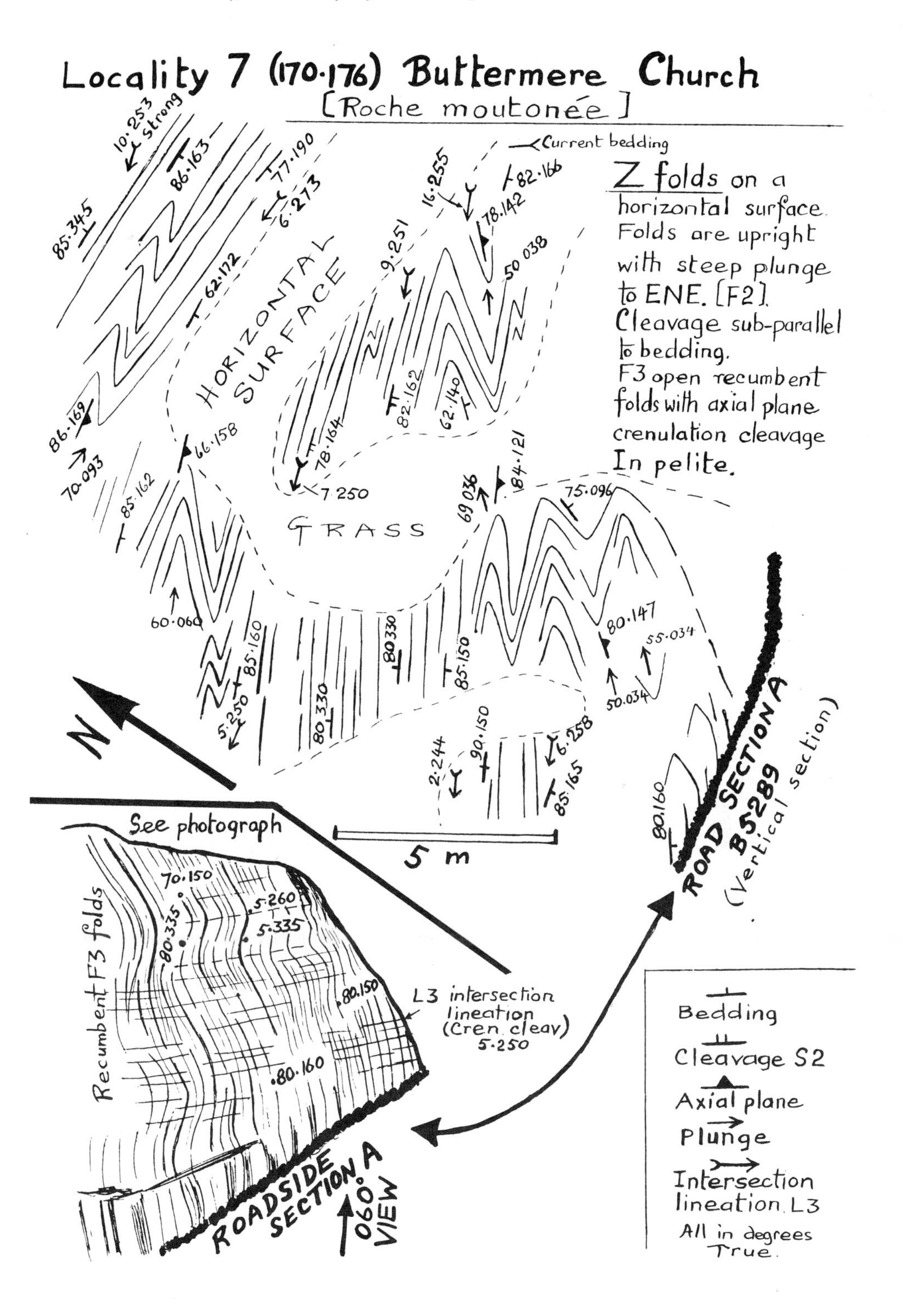
Locality 7 (170·176) Buttermere Church
[Roche moutonée]
Current bedding
Z folds on a horizontal surface. Folds are upright with steep plunge to ENE. [F2]. Cleavage sub-parallel to bedding. F3 open recumbent folds with axial plane crenulation cleavage In pelite.
HORIZONTAL SURFACE
GRASS
5 m
N
See photograph
Recumbent F3 folds
L3 intersection lineation (Cren. cleav) 5·250
ROADSIDE SECTION A
060° VIEW
ROAD SECTION A B5289 (Vertical section)
Bedding
Cleavage S2
Axial plane
Plunge
Intersection lineation L3
All in degrees True.

Rodding, mullion structures and boudins. These structures are especially common in some of the more competent schistose rocks, and, like mineral lineations, they are often parallel to fold axes. Orientations should always be recorded.

Deformed structures. There is a whole range of formerly equidimensional or symmetrical structures which can be deformed into linear elements. Pebbles of a conglomerate and reduction spots in slate are two examples.

Analysis

Analysis of structures measured in the field is usually made in the laboratory after completion of the fieldwork. However, it is important to keep a 'running plot' of measurements during fieldwork, since this will inform the surveyor whether enough measurements have been taken on particular structures, and so on. The results are then plotted stereographically for convenience of analysis.

References

Abbotts, I. L. (1978). High-potassium granites in the Masirah Ophiolite of Oman. *Geol. Mag.* **115,** 415–25.

— (1979). Intrusive processes at Ocean ridges. Evidence from the sheeted dyke complex of Masirah, Oman. *Tectonophysics* **60,** 217–33.

Allum, J. A. E. (1966). *Photogeology and regional mapping.* Oxford: Pergamon Press.

Aveline, W. T., Hughes, T. McK. and Strahan, A. (1888). The geology of the country around Kendal, Sedbergh, Bowness and Tebay. *Mem. geol. Surv. U.K.*

Boulton, G. S. (1972). Modern arctic glaciers as depositional models for former ice sheets. *J. geol. Soc. Lond.* **128,** 361–93.

Brown, P. E. (1973). A layered plutonic complex of alkali basalt parentage: the Lilloise intrusion, East Greenland. *J. geol. Soc. Lond.* **129,** 405–18.

Burgess, I. C. and Wadge, A. J. (1974). *The geology of the Cross Fell area.* London: HMSO.

Compton, R. R. (1962). *Manual of field geology.* New York: Wiley.

Doughty, P. S. (1968). Joint densities and their relation to lithology in the Great Scar Limestone. *Proc. Yorks. geol. Soc.* **36,** 479–512.

Firman, R. J. (1960). The relationship between joints and fault patterns in the Eskdale Granite and adjacent Borrowdale Volcanic Series. *Q. J. geol. Soc. Lond.* **116,** 317–40.

Gass, I. G. and Gibson, I. L. (1969). The structural evolution of the rift zones in the Middle East. *Nature, Lond.* **221,** 926–30.

Glennie, K. W., Boeuf, M. G. A., Hughes-Clark, M. W., Moody-Stuart, M., Pilaar, W. F. H. and Reinhardt, B. M. (1974). *Geology of the Oman Mountains.* Verhandelingen van het Koninklijk Nederlands geologisch Mijnbouwkunig Genootschap.

Greenwood, J. E. G. W. (1966). Photogeological maps of Western Aden Protectorate. 1:250,000. London: Directorate of Overseas Surveys.

— and Bleackley, D. (1967). Geology of the Arabian Peninsula: Aden Protectorate. *Prof. pap. U.S. geol. Surv*. 560–C.

Hamon, P. J. (1961). Manual of the stereographic projection for a geometric and kinematic analysis of folds and faults. *W. Can. res. Publ.* Series 1, No. 1.

Hancock, P. L. (1969). Jointing in the Jurassic Limestones of the Cotswold Hills. *Proc. Geol. Ass.* **80,** 219–41.

— and Atiya, M. S. (1979). Tectonic significance of mesofracture systems associated with the Lebanese segment of the Dead Sea transform fault. *J. struct. Geol*. **1,** 143–54.

— and Kadhi, A. (1978). Analysis of fractures in the Dhruma-Nisah segment of the Central Arabian graben system. *J. geol. Soc. Lond.* **135,** 339–47.
Hardie, W. G. (1968). Volcanic breccia and the Lower Old Red Sandstone unconformity, Glen Coe, Argyll. *Scott. J. Geol.* **4,** 291–9.
Hobbs, B. E., Means, W. D. and Williams, P. F. (1976). *An outline of structural geology.* New York: Wiley.
Jeans, P. J. F. (1974). The structure, metamorphism and stratigraphy of the Skiddaw Slates east of Crummock Water, Cumberland. Unpublished PhD thesis, University of Birmingham.
Lattman, L. H. and Ray, R. G. (1965). *Aerial photographs in field geology.* New York: Holt, Rinehart and Winston.
Lees, G. M. (1928). The geology and tectonics of Oman and parts of SE Arabia. *Q. J. geol. Soc. Lond.* **84,** 585–620.
Miller, V. C. (1961). *Photogeology.* New York: McGraw Hill.
Millward, D. (1980). Three ignimbrites from the Borrowdale Volcanic Group. *Proc. Yorks. geol. Soc.* **42,** 595–616.
—, Moseley, F. and Soper, N. J. (1978). The Eycott and Borrowdale Volcanic rocks. In *The Geology of the Lake District* (ed. Moseley, F.). Yorkshire Geological Society, 99–120.
Mitchell, G. H., Moseley, F., Firman, R. J., Soper, N. J., Roberts, D. E., Nutt, M. J. C. and Wadge, A. J. (1972). Excursion to the northern Lake District. *Proc. Geol. Ass.* **83,** 443–70.
Moseley, F. (1954). The Namurian of the Lancaster Fells. *Q. J. geol. Soc. Lond.* **109,** 423–54.
— (1956). The geology of the Keasden area, west of Settle, Yorkshire. *Proc. Yorks. geol. Soc.* **30,** 331–52.
— (1960). The succession and structure of the Borrowdale Volcanic rocks south-east of Ullswater. *Q.J. geol. Soc. Lond.* **116,** 55–84.
— (1962). The structure of the south-western part of the Sykes Anticline, Bowland, West Yorkshire. *Proc. Yorks geol. Soc.* **33,** 287–314.
— (1963). Fresh water in the North Libyan Desert. *R. Engrs J.* **77,** 130–40.
— (1965). Plateau calcrete, calcreted gravels, cemented dunes and related deposits of the Mallegh-Bomba region of Libya. *Ann. Geomorph.* **9,** 166–85.
— (1966). Exploration for water in the Aden Protectorate. *R. Engrs J.* **80,** 124–42.
— (1968a). Joints and other structures in the Silurian rocks of the southern Shap Fells, Westmorland. *Geol. J.* **6,** 79–96.
— (1968b). Conical folding and fracture patterns in the Pre-Betic of SE Spain. *Geol. J.* **6,** 97–104.
— (1969a). The Upper Cretaceous ophiolite complex of Masirah Island, Oman. *Geol. J.* **6,** 293–306.
— (1969b). The Aden Traps of Dhala, Musaymir and Radfan, South Yemen. *Bull. Volc.* **33,** 889–909.
—(1971a). Problems of water supply, development and use in Audhali, Dathina and eastern Fadhli, Southern Yemen. *Overseas geol. min. Resources,* **10,** 309–27.
— (1971b). A reconnaissance of the Wadi Beihan, South Yemen. *Proc. Geol. Ass.* **82,** 61–70.

— (1972a). A tectonic history of north-west England. *J. geol. Soc. Lond.* **128,** 561–98.

— (1972b). Stereoscopic ground photographs in field geology. *Mercian Geologist* **4,** 97–9.

— (1973). Diapiric and gravity tectonics in the Pre-Betic (Sierra Bernia) of south-east Spain. *Bol. geol. y min. España* **84,** 114–26.

— (1976). The Neogene and Quaternary of Akrotiri, Cyprus. *Mercian Geologist* **6,** 49–58.

— (1977). Explosion breccias in the Borrowdale volcanics of High Rigg, near Keswick, Cumbria. *Proc. Cumb. geol. Soc.* **3,** 197–207.

— (1980). The Caledonides of Northern England. In *Introduction to general geology*. (ed. Owen, T. R.). 26th Int. geol. Cong., Paris, 30–34.

— and Abbotts, I. L. (1979). The ophiolite mélange of Masirah, Oman. *J. geol. Soc. Lond.* **136,** 713–24.

— and Ahmed, S. M. (1967). Carboniferous joints in the north of England and their relation to earlier and later structures. *Proc. Yorks. geol. Soc.* **36,** 61–90.

— and Ahmed, S. M. (1973). Relationship between joints in Pre-Cambrian, Lower Palaeozoic and Carboniferous rocks in the West Midlands of England. *Proc. Yorks. geol. Soc.* **39,** 295–314.

— and Cruse, P. K. (1969). Exploitation of the main water table of NE Libya. *R. Engrs J.* **69,** 12–23.

Norman, J. W. (1968). Photogeology of linear features in areas covered with superficial deposits. *Trans. Inst. Min. Metall.* **77,** 60–77.

— (1970). Photo interpretation of boulder clay areas as an aid to engineering geological studies. *Q. J. Eng. Geol.* **2,** 149–57.

Oliver, R. L. (1961). The Borrowdale volcanic and associated rocks of the Scafell area, English Lake District. *Q. J. geol. Soc. Lond.* **117,** 377–417.

Price, N. J. (1966). *Fault and joint development in brittle and semi-brittle rock*. Oxford: Pergamon Press.

Ramsay, J. G. (1958). Superimposed folding at Loch Monar, Inverness-shire and Ross-shire. *Q. J. geol. Soc. Lond.* **113,** 271–307.

— (1967). *Folding and fracturing of rocks.* New York: McGraw-Hill.

Roberts, D. E. (1971). Structures of the Skiddaw Slates in the Caldew Valley, Cumberland. *Geol. J.* **7,** 225–38.

— (1973). Skiddaw Slate structures of the Blencathra region, Cumbria. Unpublished PhD thesis, University of Birmingham.

Rios, J. M., Villalón, C., Trigueros, E. and Navarro, A. (1958). Mapa geológico de España (1:50,000). Explicación de la hoja No. 848, Altea. Madrid: Instituto Geológico y Minero de España.

Shapland, P. C., Little, J. S. and Moseley, F. (1967). Well drilling in the Federation of South Arabia. *R. Engrs J.* **81,** 240–54.

Sheraton, J. W., Tarney, J., Wheatley, T. J. & Wright, A. E. (1973). Structural history of the Assynt District. In *The early Pre-Cambrian of Scotland and related rocks in Greenland* (ed. Park, R. G. and Tarney, J.). Geology Department, Keele University, England, 31–44.

Shotton, F. W. (1935). The stratigraphy and tectonics of the Cross Fell Inlier. *Q. J. geol. Soc. Lond.* **91,** 639–704.

— (1946). Water supply in the Middle East Campaigns. I. The main water table of the Miocene Limestone in the coastal desert of Egypt. *Wat. & Wat. Engng.* **49,** 218–26.

Soper, N. J. (1970). Three critical localities on the junction of the Borrowdale Volcanic rocks with the Skiddaw Slates in the Lake District. *Proc. Yorks. geol. Soc.* **37,** 461–93.

— and Moseley, F. (1978). Structure. In *The geology of the Lake District* (ed. Moseley, F.). Yorkshire Geological Society, 45–67.

Stauffer, M. R. (1964). The geometry of conical folds. *N.Z. J. Geol. Geophys.* **7,** 340–7.

Wadge, A. J., Harding, R. R. and Darbyshire, D. P. F. (1974). The rubidium strontium age and field relations of the Threlkeld Microgranite. *Proc. Yorks. geol. Soc.* **40,** 211–22.

Webb, B. C. (1972). N–S trending pre-cleavage folds in the Skiddaw Slate Group of the English Lake District. *Nature, phys. Sci.* **235,** 138–40.

Yates, E. M. and Moseley, F. (1967). A contribution to the glacial geomorphology of the Cheshire Plain. *Trans. Inst. Br. Geog.* **42,** 107–25.

Index